Modern Organic Synthesis

Modern Organic Synthesis

Editor: Norman Henderson

NY RESEARCH
P R E S S

New York

Published by NY Research Press
118-35 Queens Blvd., Suite 400,
Forest Hills, NY 11375, USA
www.nyresearchpress.com

Modern Organic Synthesis
Edited by Norman Henderson

International Standard Book Number: 978-1-63238-587-1 (Hardback)

Cataloging-in-Publication Data

Modern organic synthesis / edited by Norman Henderson.
   p. cm.
Includes bibliographical references and index.
ISBN 978-1-63238-587-1
1. Organic compounds--Synthesis. I. Henderson, Norman.
QD262 .M63 2018
547.2--dc23

Contents

Permissions

Index

Preface

Organic synthesis refers to that branch of chemical synthesis, which deals with the use of organic reactions to produce organic compounds. The three main sub-branches of organic synthesis are methodology and applications, total synthesis, and semisynthesis. Such selected concepts that redefine this field have been presented in this book. It unfolds the innovative aspects of the area, which will be crucial for the holistic understanding of the subject matter. This textbook, with its detailed analyses and data, will prove immensely beneficial to students involved in this area at various levels.

Given below is the chapter wise description of the book:

Chapter 1- Organic synthesis studies the creation of compounds which contain carbon through organic reactions. Organic compound are much more complex in nature as compared to inorganic compounds. Organic synthesis is a vast field with topics like semisynthesis and total synthesis. This chapter is an overview of the subject matter incorporating all the major aspects of the subject matter.

Chapter 2- Naturally occurring compounds are helpful in the development of synthetic targets that can initialize a biological process in the host. Pharmaceuticals are an important sector of the application of organic synthesis of compounds. The organic structure of a drug is important to understand its effects on a patient. This chapter has been carefully written to provide an easy understanding of the varied facets of naturally occurring compounds.

Chapter 3- The chapter strategically encompasses and incorporates the major components and key concepts of base catalyzed carbon-carbon bond formation, providing a complete understanding. The formation of a base catalyzed carbon-carbon bond depends on the bond-bond carbon formation derived from organometallic reagents. This chapter discusses the subject matter in a critical manner.

Chapter 4- This section focuses on the formation of carbon-nitrogen bonds. This bond can be categorized into the reaction caused by nucleophilic nitrogen with electrophilic carbon and the reaction of electrophilic nitrogen with nucleophilic carbon. The aspects elucidated in this chapter are of vital importance, and provide a better understanding of carbon-nitrogen bonds formation.

Chapter 5- When the carbon skeleton of a molecule has been rearranged to give a structural isomer of the original molecule, a molecular rearrangement is said to have taken place. The major components of organic synthesis are discussed in this chapter.

Indeed, my job was extremely crucial and challenging as I had to ensure that every chapter is informative and structured in a student-friendly manner. I am thankful for the support provided by my family and colleagues during the completion of this book.

Editor

Understanding Organic Synthesis

Organic synthesis studies the creation of compounds which contain carbon through organic reactions. Organic compound are much more complex in nature as compared to inorganic compounds. Organic synthesis is a vast field with topics like semisynthesis and total synthesis. This chapter is an overview of the subject matter incorporating all the major aspects of the subject matter.

Organic Synthesis

Organic synthesis is a special branch of chemical synthesis and is concerned with the construction of organic compounds via organic reactions. Organic molecules often contain a higher level of complexity than purely inorganic compounds, so that the synthesis of organic compounds has developed into one of the most important branches of organic chemistry. There are several main areas of research within the general area of organic synthesis: *total synthesis*, *semisynthesis*, and *methodology*.

Total Synthesis

A total synthesis is the complete chemical synthesis of complex organic molecules from simple, commercially available (petrochemical) or natural precursors. Total synthesis may be accomplished either via a linear or convergent approach. In a *linear* synthesis—often adequate for simple structures—several steps are performed one after another until the molecule is complete; the chemical compounds made in each step are called *synthetic intermediates*. For more complex molecules, a convergent synthetic approach may be preferable, one that involves individual preparation of several "pieces" (key intermediates), which are then combined to form the desired product.

Robert Burns Woodward, who received the 1965 Nobel Prize for Chemistry for several total syntheses (e.g., his 1954 synthesis of strychnine), is regarded as the father of modern organic synthesis. Some latter-day examples include Wender's, Holton's, Nicolaou's, and Danishefsky's total syntheses of the anti-cancer therapeutic, paclitaxel (trade name, Taxol).

Methodology and Applications

Each step of a synthesis involves a chemical reaction, and reagents and conditions for each of these reactions must be designed to give an adequate yield of pure product, with as little work as possible. A method may already exist in the literature for making one of the early synthetic intermediates, and this method will usually be used rather than an effort to "reinvent the wheel". However, most intermediates are compounds that have never been made before, and these will normally be made using general methods developed by methodology researchers. To be useful, these methods need to give high yields, and to be reliable for a broad range of substrates. For practical applications, additional hurdles include industrial standards of safety and purity.

Methodology research usually involves three main stages: *discovery*, *optimisation*, and studies of *scope and limitations*. The *discovery* requires extensive knowledge of and experience with chemical reactivities of appropriate reagents. *Optimisation* is a process in which one or two starting compounds are tested in the reaction under a wide variety of conditions of temperature, solvent, reaction time, etc., until the optimum conditions for product yield and purity are found. Finally, the researcher tries to extend the method to a broad range of different starting materials, to find the scope and limitations. Total syntheses are sometimes used to showcase the new methodology and demonstrate its value in a real-world application. Such applications involve major industries focused especially on polymers (and plastics) and pharmaceuticals.

Stereoselective Synthesis

Most complex natural products are chiral, and the bioactivity of chiral molecules varies with the enantiomer. Historically, total syntheses targeted racemic mixtures, mixtures of both possible enantiomers, after which the racemic mixture might then be separated via chiral resolution.

In the later half of the twentieth century, chemists began to develop methods of stereoselective catalysis and kinetic resolution whereby reactions could be directed to produce only one enantiomer rather than a racemic mixture. Early examples include stereoselective hydrogenations (e.g., as reported by William Knowles and Ryōji Noyori), and functional group modifications such as the asymmetric epoxidation of Barry Sharpless; for these specific achievements, these workers were awarded the Nobel Prize in Chemistry in 2001. Such reactions gave chemists a much wider choice of enantiomerically pure molecules to start from, where previously only natural starting materials could be used. Using techniques pioneered by Robert B. Woodward and new developments in synthetic methodology, chemists became more able to take simple molecules through to more complex molecules without unwanted racemisation, by understanding stereocontrol, allowing final target molecules to be synthesised pure enantiomers (i.e., without need for resolution). Such techniques are referred to as *stereoselective synthesis*.

Synthesis Design

Elias James Corey brought a more formal approach to synthesis design, based on retrosynthetic analysis, for which he won the Nobel Prize for Chemistry in 1990. In this approach, the synthesis is planned backwards from the product, using standard rules. The steps "breaking down" the parent structure into achievable component parts are shown in a graphical scheme that uses *retrosynthetic arrows* (drawn as ⇒, which in effect, mean "is made from").

More recently, and less widely accepted, computer programs have been written for designing a synthesis based on sequences of generic "half-reactions".

Inorganic Compound

A chemical compound is termed inorganic if it fulfills one or more of the following criteria:

- There is an absence of carbon in its composition

- It is of a non-biologic origin

- It cannot be found or incorporated into a living organism

There is no clear or universally agreed-upon distinction between organic and inorganic compounds. Organic chemists traditionally and generally refer to any molecule containing carbon as an organic compound and by default this means that inorganic chemistry deals with molecules lacking carbon. As many minerals are of biological origin, biologists may distinguish organic from inorganic compounds in a different way that does not hinge on the presence of a carbon atom. Pools of organic matter, for example, that have been metabolically incorporated into living tissues persist in decomposing tissues, but as molecules become oxidized into the open environment, such as atmospheric CO_2, this creates a separate pool of inorganic compounds. The International Union of Pure and Applied Chemistry, an agency widely recognized for defining chemical terms, does not offer definitions of inorganic or organic compounds. Hence, the definition for an inorganic versus an organic compound in a multidisciplinary context spans the division between organic life living (or animate) and inorganic non-living (or inanimate) matter. In broader speech, the term commonly referred to compounds synthesised by purely geological systems, in contrast to those with a biological component in their origin.

Traditional Usage

The Wöhler synthesis is the conversion of ammonium cyanate into urea. This chemical reaction was discovered in 1828 by Friedrich Wöhler and is considered the starting point of modern organic chemistry.

The Wöhler synthesis is of great historical significance because for the first time an organic compound was produced from inorganic reactants. This finding went against the mainstream theory of that time called vitalism, which stated that organic matter possessed a special force or *vital force* inherent to all things living. For this reason a sharp boundary existed between organic and inorganic compounds. Urea was discovered in 1799 and could until 1828 only be obtained from biological sources such as urine. Wöhler reported to his mentor Berzelius:

"I cannot, so to say, hold my chemical water and must tell you that I can make urea without thereby needing to have kidneys, or anyhow, an animal, be it human or dog".

Modern Usage

Inorganic compounds can be defined as any compound that is not organic compound. Some simple compounds which contain carbon are usually considered inorganic. These include carbon monoxide, carbon dioxide, carbonates, cyanides, cyanates, carbides, and thiocyanates. Many of these are normal parts of mostly organic systems, including organisms, which means that describing a chemical as inorganic does not obligately mean that it does not occur within living things. In contrast, methane and formic acid are generally considered to be simple examples of organic compounds, although the Inorganic Crystal Structure Database (ICSD), in its definition of "inorganic" carbon compounds, states that such compounds may contain *either* C-H or C-C bonds, but not both.

Coordination Chemistry

A large class of compounds discussed in inorganic chemistry textbooks are coordination compounds. Examples range from substances that are strictly inorganic, such as $[Co(NH_3)_6]Cl_3$, to organometallic compounds, such as $Fe(C_5H_5)_2$, and extending to bioinorganic compounds, such as the hydrogenase enzymes.

Mineralogy

Minerals are mainly oxides and sulfides, which are strictly inorganic, although they may be of biological origin. In fact, most of the Earth is inorganic. Although the components of Earth's crust are well-elucidated, the processes of mineralization and the composition of the deep mantle remain active areas of investigation, which are covered mainly in geology-oriented venues.

Organic Compound

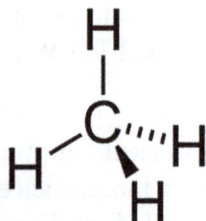

Methane, CH_4; it is one of the simplest organic compounds.

An organic compound is virtually any chemical compound that contains carbon, although a consensus definition remains elusive and likely arbitrary. Organic compounds are rare terrestrially, but of central importance because all known life is based on organic compounds. The most basic petrochemicals are considered the building blocks of organic chemistry.

Definitions of Organic vs Inorganic

For historical reasons discussed below, a few types of carbon-containing compounds, such as carbides, carbonates, simple oxides of carbon (for example, CO and CO_2), and cyanides are considered inorganic. The distinction between *organic and inorganic* carbon compounds, while "useful in organizing the vast subject of chemistry... is somewhat arbitrary".

Organic chemistry is the science concerned with all aspects of organic compounds. Organic synthesis is the methodology of their preparation.

History

Vitalism

The word *organic* is historical, dating to the 1st century. For many centuries, Western alchemists believed in vitalism. This is the theory that certain compounds could be synthesized only from their classical elements—earth, water, air, and fire—by the action of a "life-force" (*vis vitalis*) that only organisms possessed. Vitalism taught that these "organic" compounds were fundamentally different from the "inorganic" compounds that could be obtained from the elements by chemical manipulation.

Vitalism survived for a while even after the rise of modern atomic theory and the replacement of the Aristotelian elements by those we know today. It first came under question in 1824, when Friedrich Wöhler synthesized oxalic acid, a compound known to occur only in living organisms, from cyanogen. A more decisive experiment was Wöhler's 1828 synthesis of urea from the inorganic salts potassium cyanate and ammonium sulfate. Urea had long been considered an "organic" compound, as it was known to occur only in the urine of living organisms. Wöhler's experiments were followed by many others, in which increasingly complex "organic" substances were produced from "inorganic" ones without the involvement of any living organism.

Modern Classification

Even though vitalism has been discredited, scientific nomenclature retains the distinction between *organic* and *inorganic* compounds. The modern meaning of *organic compound* is any compound that contains a significant amount of carbon—even though many of the organic compounds known today have no connection to any substance found in living organisms.

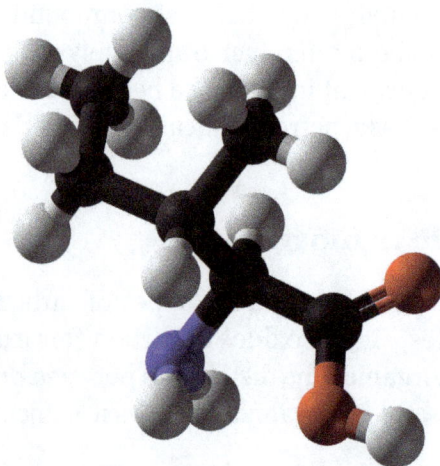

The L-isoleucine molecule, $C_6H_{13}NO_2$, showing features typical of organic compounds. Carbon atoms are in black, hydrogens gray, oxygens red, and nitrogen blue.

The organic compound L-isoleucine molecule presents some features typical of organic compounds: carbon–carbon bonds, carbon–hydrogen bonds, as well as covalent bonds between carbon to oxygen and to nitrogen.

Still, even the broadest definition (of "carbon-containing molecules" as organic) requires excluding alloys that contain carbon, including steel. Other 'excluded' materials are: compounds such as carbonates and carbonyls, simple oxides of carbon, simple carbon halides and sulfides, the allotropes of carbon, and cyanides not containing the $-C\equiv N$ functional group—all which are considered inorganic.

The "C-H" definition excludes compounds that are (historically and practically) considered organic. Neither urea nor oxalic acid is organic by this definition, yet they were two key compounds in the vitalism debate. The IUPAC Blue Book on organic nomenclature specifically mentions urea and oxalic acid. Other compounds lacking C-H bonds but traditionally considered organic include benzenehexol, mesoxalic acid, and carbon tetrachloride. Mellitic acid, which contains no C-H bonds, is considered a possible organic substance in Martian soil.

The "C-H bond-only" rule also leads to somewhat arbitrary divisions in sets of carbon-fluorine compounds. For example, CF_4 would be considered by this rule to be "inorganic", whereas CF_3H would be organic.

Classification

Organic compounds may be classified in a variety of ways. One major distinction is between natural and synthetic compounds. Organic compounds can also be classified or subdivided by the presence of heteroatoms, e.g., organometallic compounds, which feature bonds between carbon and a metal, and organophosphorus compounds, which feature bonds between carbon and a phosphorus.

Another distinction, based on the size of organic compounds, distinguishes between small molecules and polymers.

Natural Compounds

Natural compounds refer to those that are produced by plants or animals. Many of these are still extracted from natural sources because they would be more expensive to produce artificially. Examples include most sugars, some alkaloids and terpenoids, certain nutrients such as vitamin B_{12}, and, in general, those natural products with large or stereoisometrically complicated molecules present in reasonable concentrations in living organisms.

Further compounds of prime importance in biochemistry are antigens, carbohydrates, enzymes, hormones, lipids and fatty acids, neurotransmitters, nucleic acids, proteins, peptides and amino acids, lectins, vitamins, and fats and oils.

Synthetic Compounds

Compounds that are prepared by reaction of other compounds are known as "synthetic". They may be either compounds that already are found in plants or animals or those that do not occur naturally.

Most polymers (a category that includes all plastics and rubbers), are organic synthetic or semi-synthetic compounds.

Biotechnology

Many organic compounds—two examples are ethanol and insulin—are manufactured industrially using organisms such as bacteria and yeast. Typically, the DNA of an organism is altered to express compounds not ordinarily produced by the organism. Many such biotechnology-engineered compounds did not previously exist in nature.

Databases

- The *CAS* database is the most comprehensive repository for data on organic compounds. The search tool *SciFinder* is offered.

- The *Beilstein database* contains information on 9.8 million substances, covers the scientific literature from 1771 to the present, and is today accessible via Reaxys. Structures and a large diversity of physical and chemical properties is available for each substance, with reference to original literature.

- *PubChem* contains 18.4 million entries on compounds and especially covers the field of medicinal chemistry.

A great number of more specialized databases exist for diverse branches of organic chemistry.

Structure Determination

The main tools are proton and carbon-13 NMR spectroscopy, IR Spectroscopy, Mass spectrometry, UV/Vis Spectroscopy and X-ray crystallography.

Phosphorus-Containing Compounds

The phosphorus reagents have three characteristics: the ease with which phosphorus (III) is converted into phosphorus (V); the relatively strong bonds formed between phosphorus and oxygen; and the availability of vacant 3d orbitals for bonding.

Wittig Olefination

The Wittig reaction for the synthesis of alkenes stems from two properties: First, it is specific for the conversion of aldehydes or ketone to alkenes. Second, the carbonyl compounds can contain a variety of other functional groups.

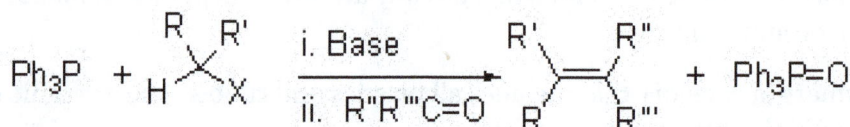

Mechanism

The phosphorus ylides are prepared by quaternizing a tervalent phosphorus compounds with an alkyl halide and treating the salt with base. The phosphorane adds as a carbon nucleophile to the carbonyl group and the resulting intermediate reacts via a cyclic intermediate to form the alkene.

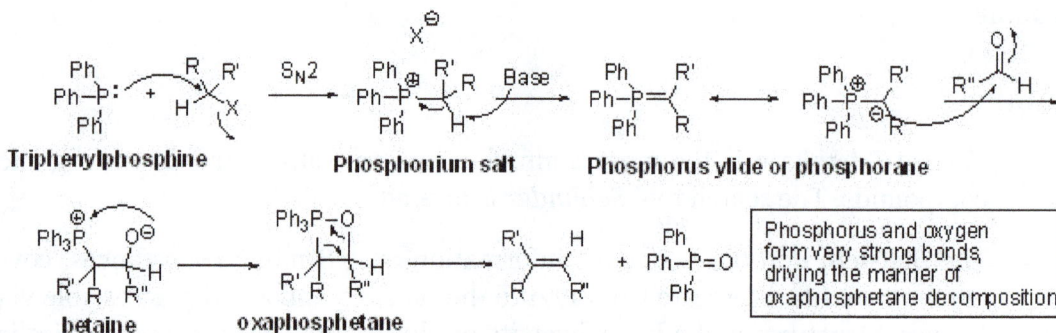

Notes:

Other phosphines may be used for this reaction, but the choice should not contain a proton that could be abstracted by base, because a mixture of desired and undesired ylides would be formed.

Usually strong bases such as BuLi, NaH and NaNH$_2$ are used.

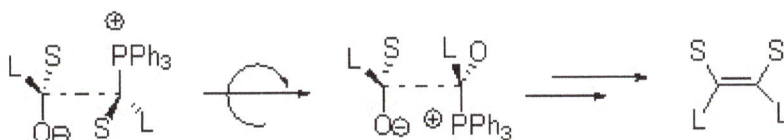

| Prefered anti attack of ylide, minimizing steric hinderance | Bond rotation follows to form the betaine | the result often gives the z-alkene |

Simple phosphoranes are very reactive and are unstable in the presence of air or moisture. They are therefore prepared in a scrupulously dry solvent under nitrogen and the carbonyl compound is added as soon as the phosphorane has been formed.

More stable phosphoranes are obtained when a −M substituent is adjacent to the anionic carbon. However, although they react with aldehydes, they do not do so effectively with ketones and for this purpose modification has been introduced in which triphenylphosphine is replaced by triethyl phosphate (Wadsworth-Emmons Reaction).

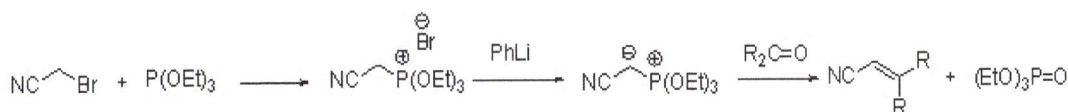

Examples

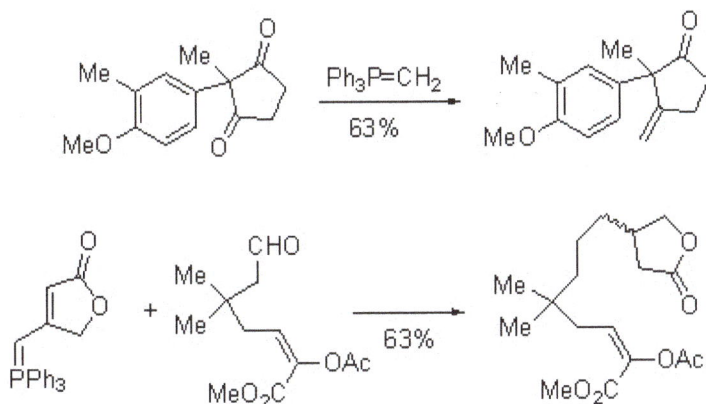

Wittig Indole Synthesis

Indole is an important structural unit present in many natural products and biologically active compounds. The reaction of (2-amidobenzyl)triphenylphosphonium salt with base allows the synthesis of 2-substituted indoles.

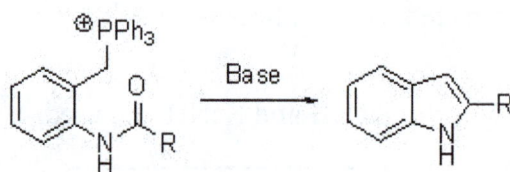

Mechanism

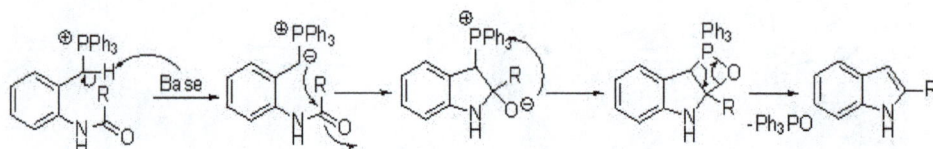

Examples

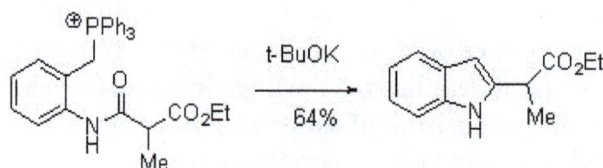

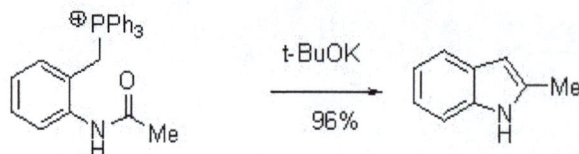

Arbuzov Reaction (Michaelis-Arbuzov Reaction)

It is an effective method for the synthesis of an alkyl phosphonate from a trialkyl phosphate and an alkyl halide. The reaction finds wide applications in the synthesis of phosphonate esters which are used in the Horner-Emmons Reaction.

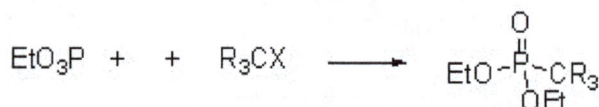

Mechanism

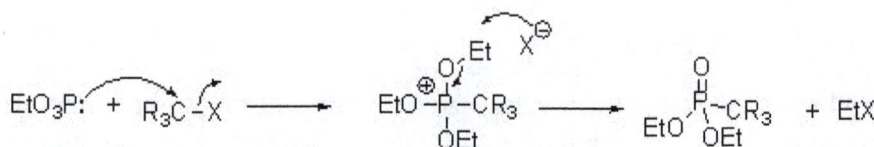

Michaelis Reaction (Michaelis-Becker Reaction)

It is an alternative method for the synthesis of alkyl phosphonate esters. The yield of the process is usually less compared to the above mentioned Michaelis-Arbuzov reaction.

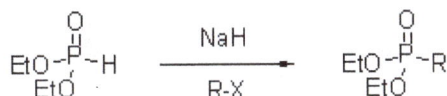

Mechanism

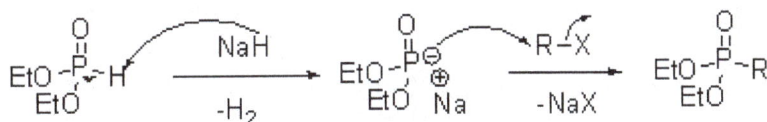

Kabachnik-Fields Reaction

The three-component reaction of a carbonyl, an amine and a hydrophosphoryl compound leads to the formation of α-aminophosphonates which is very important in drug discovery research for generating peptidomimetic compounds.

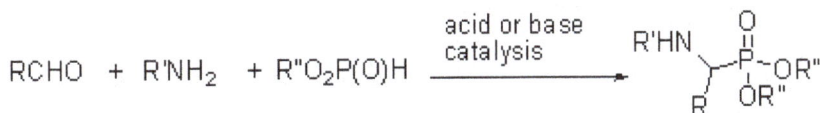

The Photo-Arbuzov Reaction

The direct UV irradiation of phosphite can give phosphonate in moderate to good yield. The reactions of both acyclic and cyclic phosphites have been explored.

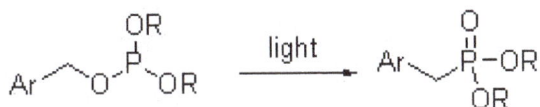

Examples

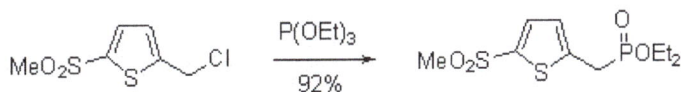

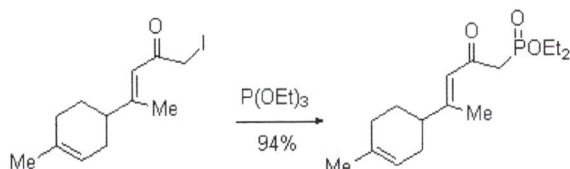

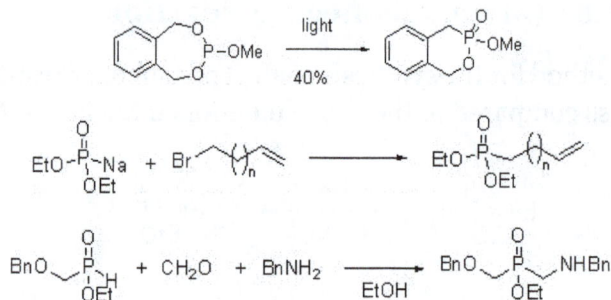

Mitsunobu Reaction

The alkoxyphosphonium ion generated from diethyl azodicarboxyliate (DEAD), triphenyl phosphine and alcohol undergoes reaction with nucleophile (usually carboxylic acid) by S$_N$2 character. The reaction exploits, first, the reactivity of the azo compound as an electrophile in the formation of the first phosphonium ion and, second, the good leaving group property of the reduced azo compound.

Mechanism

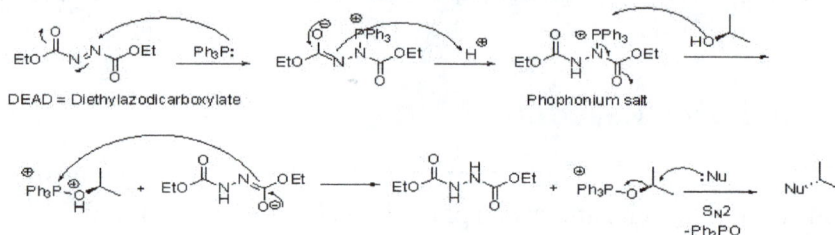

Examples

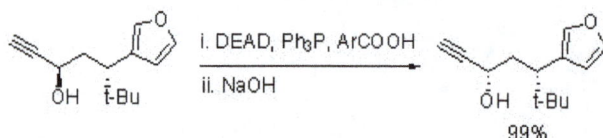

Vilsmeier-Haack Reaction

This reaction allows formylation of the reaction of activated alkenes as well as arenes.

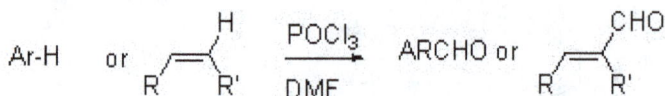

Mechanism

The formylating agent is generated in situ from DMF and POCl$_3$.

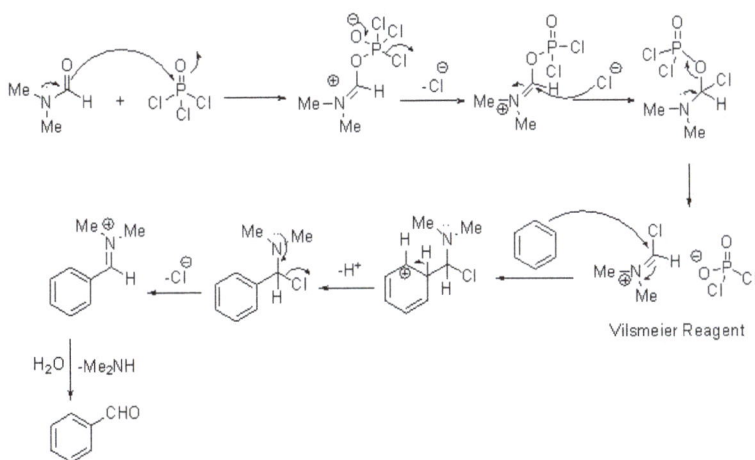

Vilsmeier Reagent

Sulfur-Containing Compounds

Sulfur-containing compounds find wide applications in organic synthesis. This section covers some of the importance processes.

Reactions with Carbonyl Compounds

Julia Olefination (Julia-Lythgoe Olefination)

Aldehydes as well as ketones undergo reaction with sulfone having α -H in the presence of base to give an adduct which could be treated with acetic anhydride to afford acetyl derivative that could proceed reductive elimination in the presence of sodium amalgam to form alkene. One of the important characteristics of this reaction is its high stereoselectivity for (E)-disubstituted alkenes.

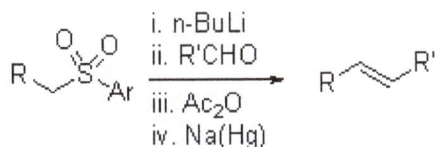

Mechanism

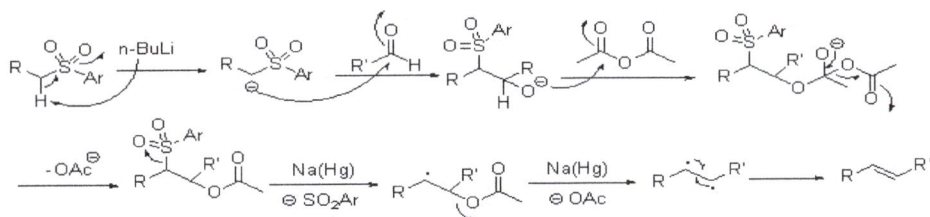

Julia Coupling

Sulfones having α -hydrogen can be coupled with aldehydes using base.

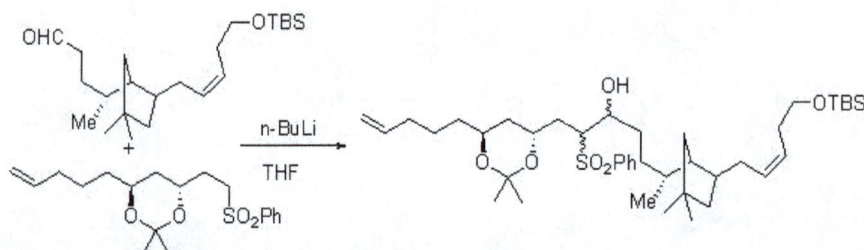

Examples

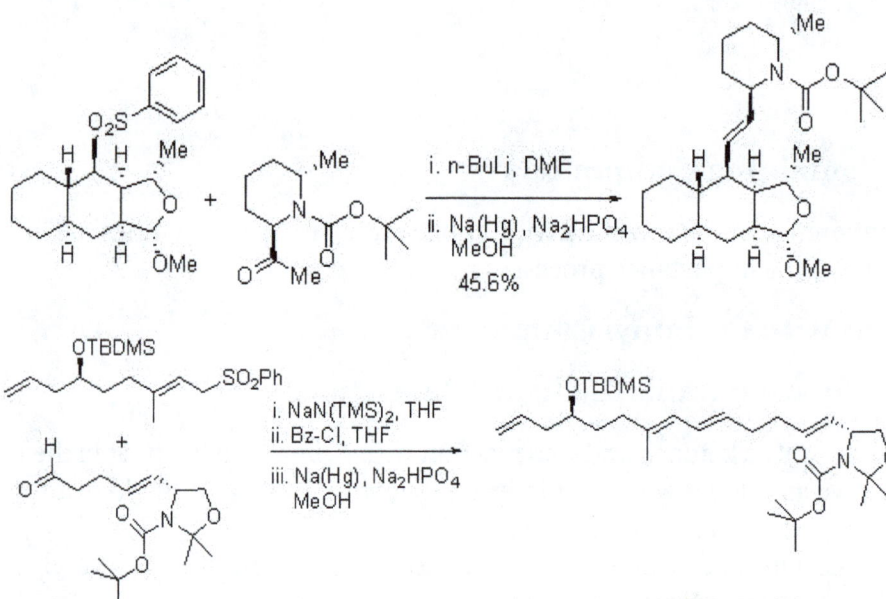

Corey-Chaykovsky Reaction

The first stage of the reaction of a sulfur ylide with an aldehydes or ketone compound consists of nucleophilic addition, the resultant adduct then proceeds intramolecular nucleophilic substitution to give an epoxide. In case of α, β -unsaturated compounds, based on the nucleophilicity of the ylide either epoxide or cyclopropanation could be formed.

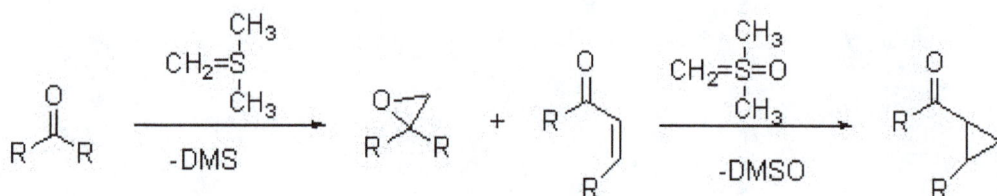

Mechanism

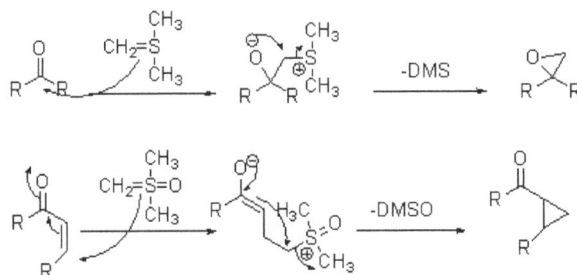

Recently, chiral version of the reaction has been extensively studied for the synthesis of optically active epoxides as the principle shown in figure. The products are usually obtained with high enantioselectivity.

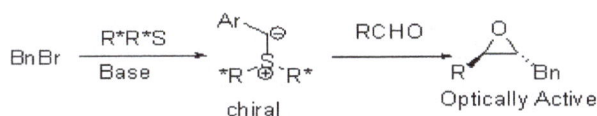

Examples

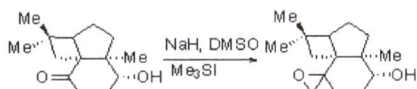

C. F. D. Amigo, I. G. Collado, J. R. Hanson, R. Hernandez-Galam, P. B. Hitchcock, A. J. Macias-Sanchez, D. J. Mobbs, M. *J. Org. Chem.* **2001**, *66*, 4327.

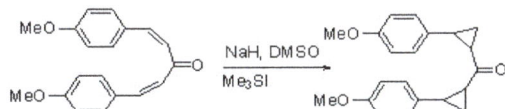

K. Hantawong, W. S. Murphy, N. Russell, D. R. Boyd, *Tetrahedron Lett.* **1984**, *25*, 999.

Rearrangement of Sulfur Ylides

Sulfur ylides rearrange as shown in figure. In case of allylic sulfur ylide, [3,2]-sigmatropic rearrangement is observed.

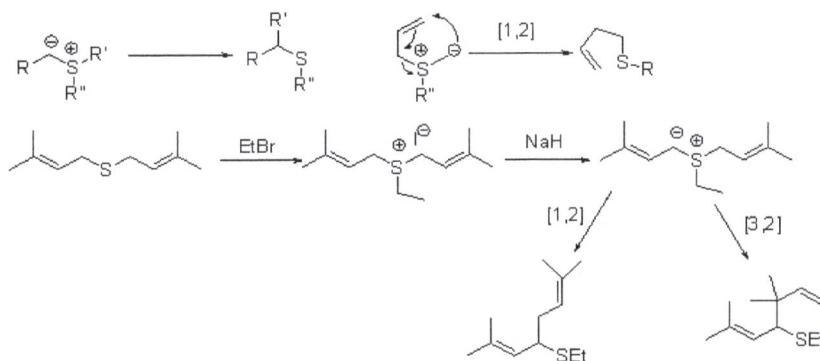

Aza- and oxa-sulfonium salts give ylides that can rearrange to give useful aromatic compounds.

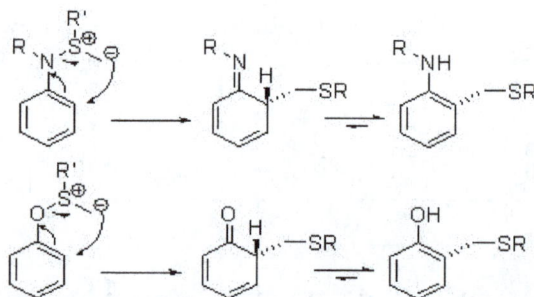

Reactions of the Dimsyl Anion

Dimsyl anion is very reactive nucleophile and can be used for various synthetic applications. The sulfur substituent can be easily removed by reduction or thermally.

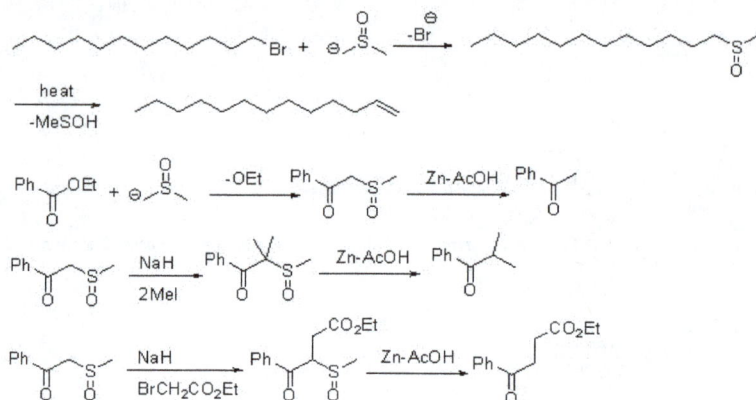

Sulfoxide Elimination

The compound containing an activated C-H bond can underogo reaction with diphenyl or dimethyl disulfide in the presence of base to give substituted sulfide that could be readily oxidized to sulfoxide. The latter readily undergoes elimination on heating to give α, β - unsaturated carbonyl compound.

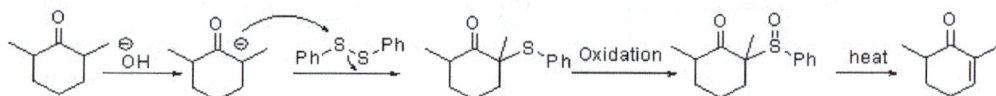

Corey-Winter Olefination

This method gives an effective route for the transformation of 1,2-diol to alkenes. The cyclic thiocarbonate formed from 1,2-diol and thiocarbonyldiimidaole undergoes reaction with phosphorus reagent via a syn elimination to afford alkene.

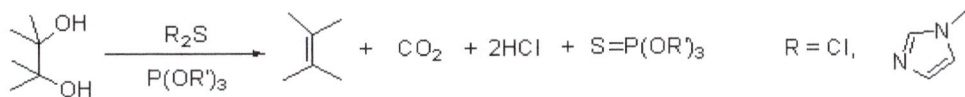

Mechanism

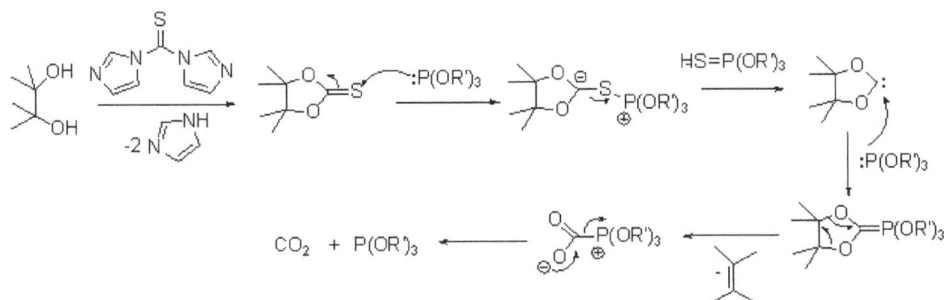

Examples

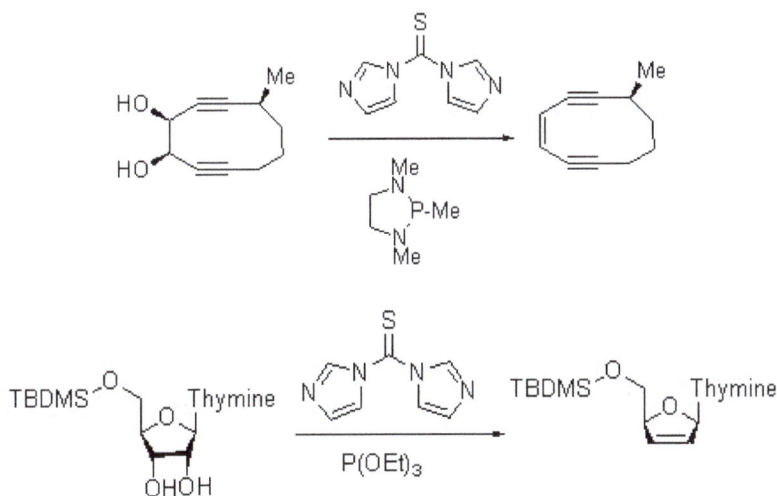

Andersen Sulfoxide Synthesis

Synthesis of chiral sulfoxide can be accomplished from sulfinyl chloride by reaction with chiral auxiliaries followed by substitution of the separated diastereomers with nucleophiles.

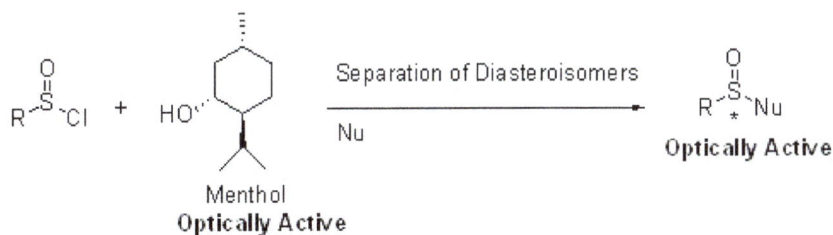

Mechanism

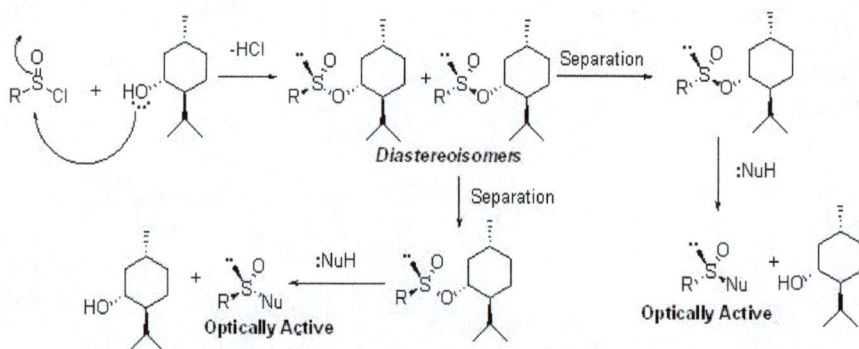

Examples

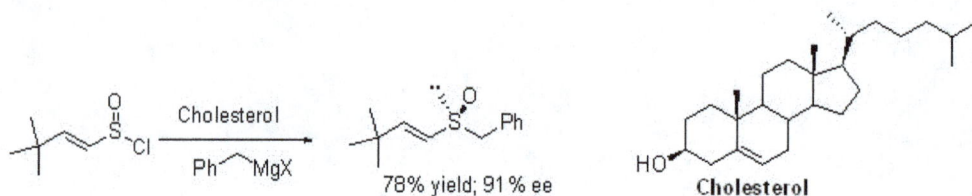

78% yield; 91 % ee

Cholesterol

Corey-Seebach Reaction

Corey-Seebach reaction provides an effective route for the transformation of aldehydes to ketones. The aldehydes can be readily reacted with thiol using acid catalysis to afford dithioacetal. The acidic hydrogen of the acetal can then be removed by base such n-Bu-Li and the carbonanion, stabilized by vacant d orbital of sulfur atom, can be alkylated in high yield. The resultant thioketal can be hydrolytically cleaved in the presence of mercury(II) salt.

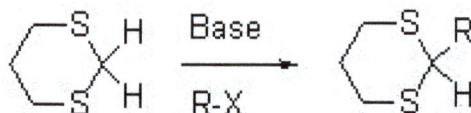

Mechanism

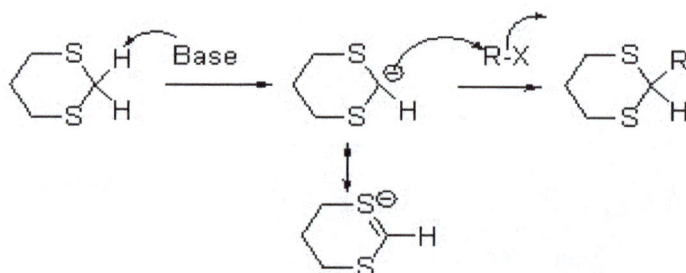

Examples

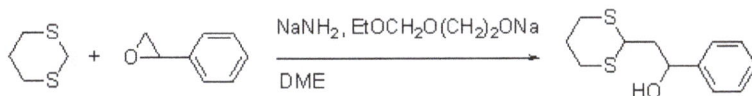

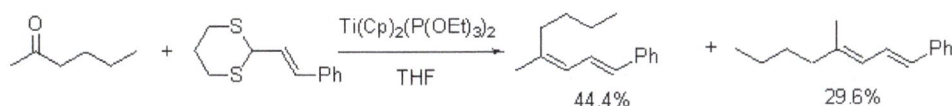

44.4%   29.6%

Corey-Nicolaou Macrocyclization

It is one of the popular macrolactonizations in organic synthesis. First, the thioester is formed from the carboxylic group and 2,2'-dipyridyl disulfide at room temperature which on reflux undergoes lactonization.

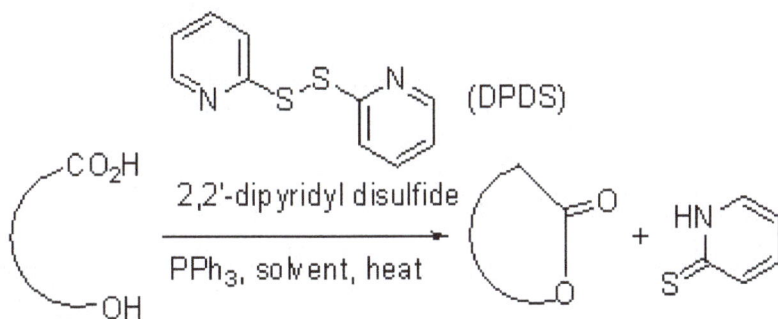

Mechanism

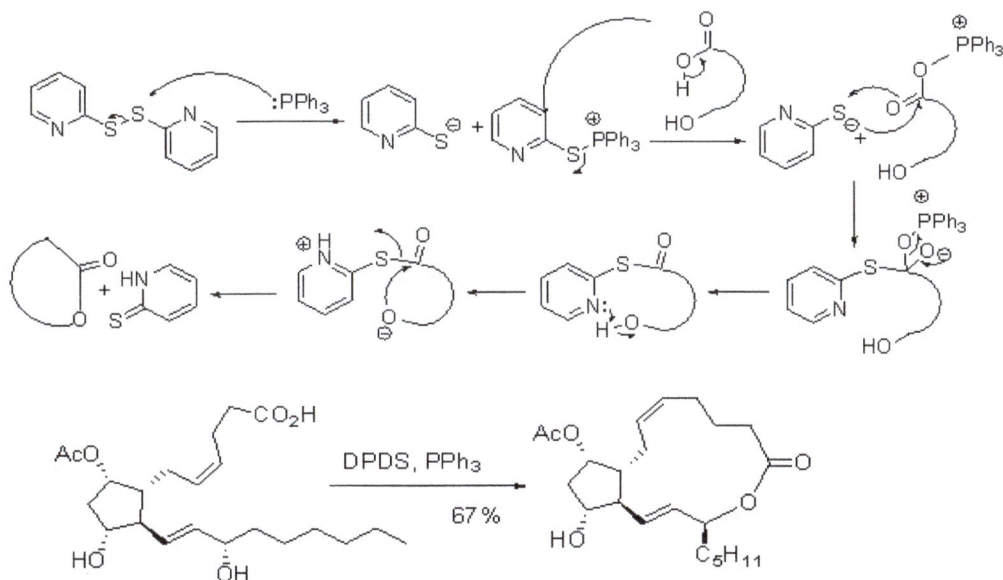

Kahne Glycosylation

It is a convenient synthesis of glycosides and disaccharides or oligosaccharides between glycosyl phenyl sulfoxide and an acceptor in the presence of a glycosylation promoter such as triflic anhydride (Tf$_2$O).

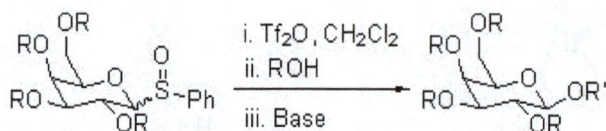

Mechanism

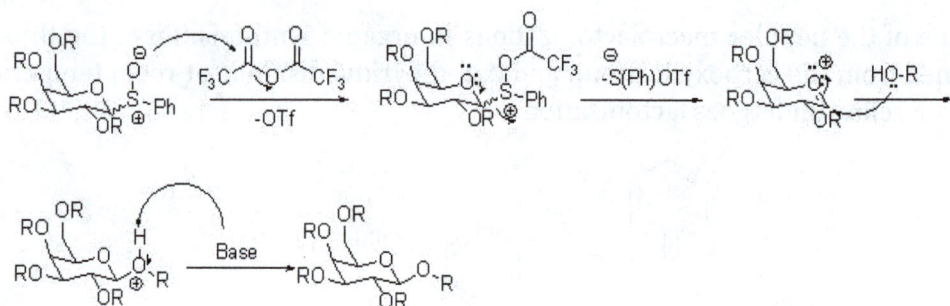

Examples

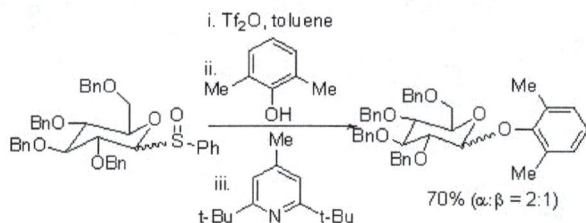

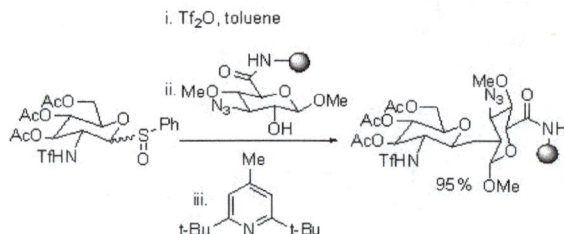

Silicon-Containing Compounds

Both silicon and carbon have similarity in having valency of four and formation of tetrahedral compounds. Regarding the differences, carbon forms many stable trigonal and linear compounds having p bonds, while silicon forms few. This is because of the strength of the silicon-oxygen σ bond (368 KJ mol^{-1}) as well as the relative weakness of the silicon-silicon (230 KJ mol^{-1}) bond.

Nucleophilic Substitution Reactions

Nucleophilic substitution at silicon differs in comparison to carbon compounds. For example, trimethylsilyl chloride does not react via S_N1 pathway which is familiar with the analogous carbon compound t-butyl chloride. This is because the S_N2 reaction at silicon is too good.

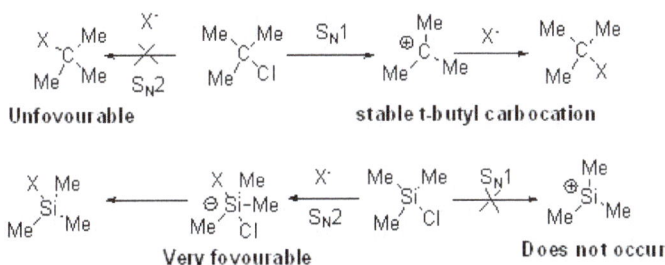

Unfovourable   stable t-butyl carbocation

Very fovourable   Does not occur

Let us compare the S_N2 reaction at silicon with the S_N2 reaction at carbon. Alkyl halides are soft electrophiles but silyl halides are hard electrophiles. The best nucleophiles for saturated carbon are neutral or based on elements down the periodic table, whereas the best nucleophiles to silicon are charged or based on highly electronegative atoms. A familiar example is the reaction of enolates at carbon with alkyl halides but at oxygen with silyl chlorides.

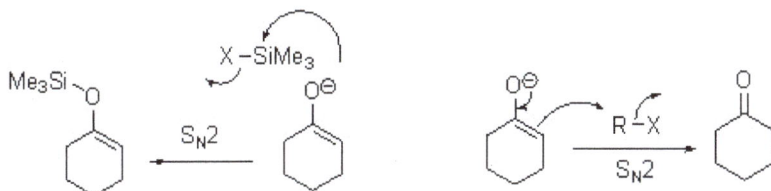

Furthermore, the S_N2 reaction at carbon is not much affected by partial positive charge (δ +) on the carbon atom. However, the S_N2 reaction at silicon is affected by the charge on silicon. For example, the most electrophilic silyl triflates react 10^9 times fast with oxygen nucleophiles than silyl chlorides do.

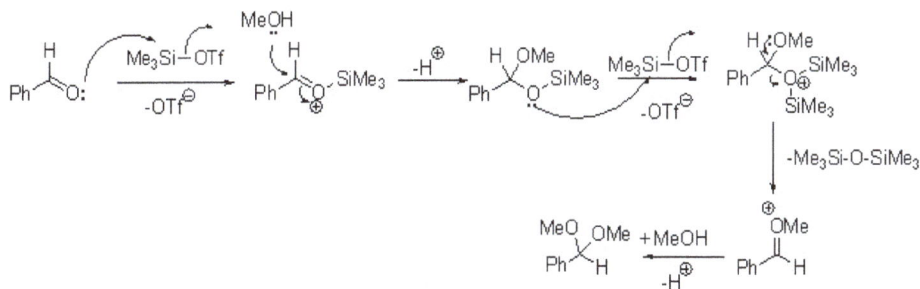

Application as Protecting Groups for Alcohols

Silicon based protecting groups are the versatile for alcohols. They can be easily introduced and removed in high yield without affecting the rest of the molecule in a wide range of conditions. The rate of the introduction as well as the removel depends on the steric nature of the alcohols as well as the silyl group.

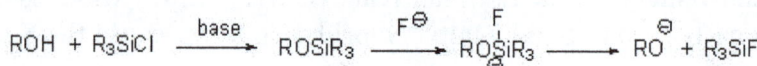

$$ROH + R_3SiCl \xrightarrow{\text{base}} ROSiR_3 \xrightarrow{F^{\ominus}} RO\overset{F}{\underset{\ominus}{Si}}R_3 \longrightarrow RO^{\ominus} + R_3SiF$$

Some the commonly used protecting for alcohols follows:

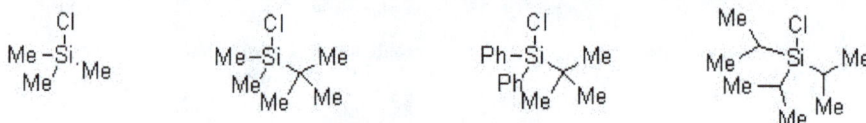

$$ \underset{Me}{\overset{Cl}{Me-Si}}Me \qquad \underset{Me}{\overset{Cl}{Me-Si}}\overset{Me}{\underset{Me}{\diagup}}Me \qquad \underset{Ph}{\overset{Cl}{Ph-Si}}\overset{Me}{\underset{Me}{\diagup}}Me \qquad \underset{Me}{\overset{Me}{\diagdown}}\underset{Me}{\overset{Cl}{Si}}\overset{Me}{\underset{Me}{\diagup}}Me $$

Application as Protecting Groups for Alkynes

The silyl group can be used to protect the terminus of the alkyne during the reaction and can also be easily removed with fluoride or sodium hydroxide.

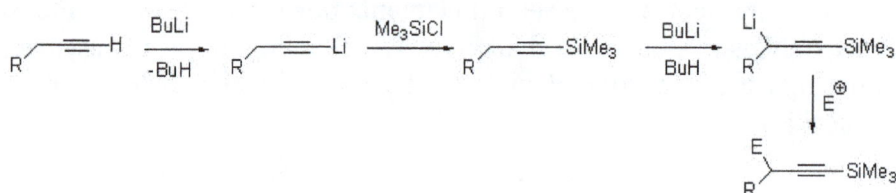

$$ R-\!\!\!\equiv\!\!\!-H \xrightarrow[-BuH]{BuLi} R-\!\!\!\equiv\!\!\!-Li \xrightarrow{Me_3SiCl} R-\!\!\!\equiv\!\!\!-SiMe_3 \xrightarrow[BuH]{BuLi} \overset{Li}{R}\!\!\!\equiv\!\!\!-SiMe_3 $$

$$ \downarrow E^{\oplus} $$

$$ \overset{E}{R}\!\!\!\equiv\!\!\!-SiMe_3 $$

Directive Influence of SiR$_3$ in Electrophilic Reactions

Alkynylsilanesoo

Similar to alkynes, silylated alkynes are too nucleophilic towards electrophiles. However, the presence of silicon has a dramatic effect on the regioselectivity of the reaction: the attack occurs only at the atom directly bonded to silicon. This is due to the stabilization of the intermediate vacant p orbital by the filled C-Si sigma orbital.

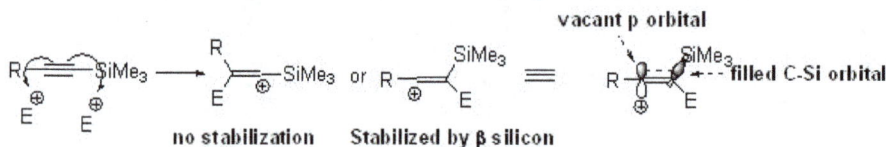

$$ R-\!\!\!\equiv\!\!\!-SiMe_3 \longrightarrow \underset{E}{\overset{R}{\diagup}}\!\!=\!\!\overset{\oplus}{}\!\!-SiMe_3 \text{ or } R-\!\!\!\underset{\oplus}{\equiv}\!\!\!-\overset{SiMe_3}{\underset{E}{}} \equiv R-\!\!\!\underset{\oplus}{}\!\!\!-\overset{SiMe_3}{\underset{E}{}} $$

vacant p orbital

filled C-Si orbital

no stabilization   Stabilized by β silicon

The stabilization of the cation weakens the C-Si bond by the delocalization of the electron density. The attack of a nucleophile on silicon readily removes it from the organic fragment and the net result is the electrophilic substitution in that the silicon is replaced by the electrophile.

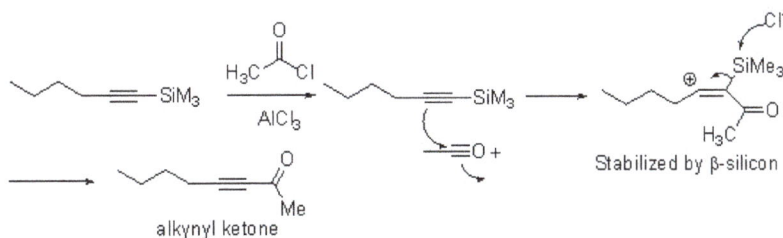

alkynyl ketone

Stabilized by β-silicon

Vinyl Silanes

The controlled reduction of alkynyl silanes can produce vinayl silanes. The stereochemistry depends on the methods used. Lindlar hydrogenation takes place via cis fashion, while red Al reduction of propargylic alcohol gives E -isomer.

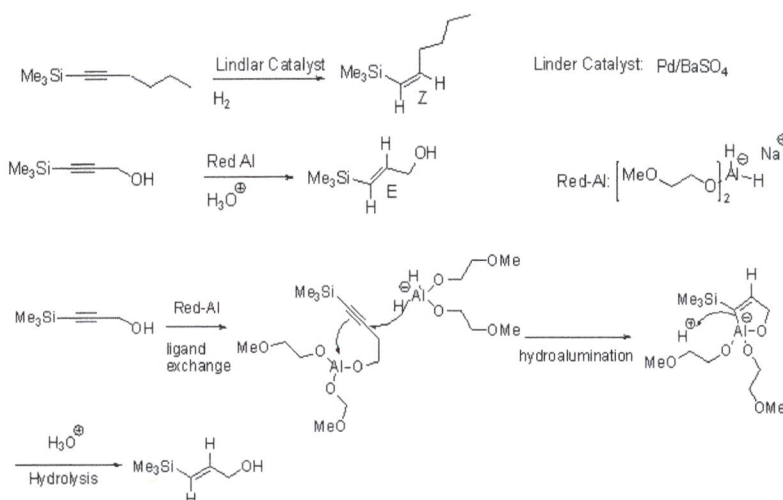

Alternatively, hydrosilylation of simple alkyne can give E-vinyl silane that could be irradiated to afford Z-isomer.

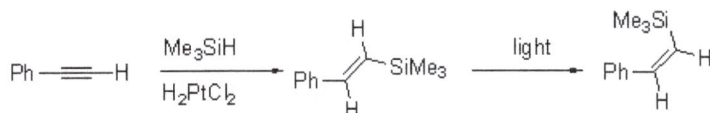

In addition, metal-halogen exchange of vinyl halide can give vinylic organometallic compound that could be cross-coupled with silyl chloride. These reactions can take place with retention of configuration.

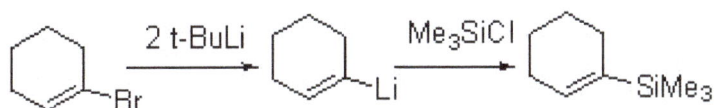

The reactions of vinyl silanes with electrophiles afford an effective route for the synthesis of alkenes with high stereoselectivity. The stereochemistry is important because the exchange usually occurs with retention of geometry.

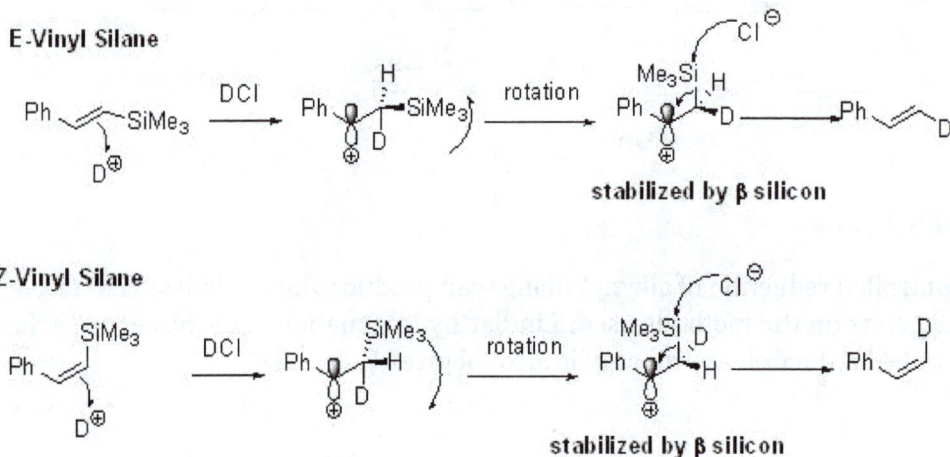

E-Vinyl Silane

stabilized by β silicon

Z-Vinyl Silane

stabilized by β silicon

Aryl Silanes

The same sort of mechanism involves for reactions of aryl silanes with electrophiles. In these reactions the silyl group is replaced by the electrophiles at the same atom on the ring. This is called as ipso substitution.

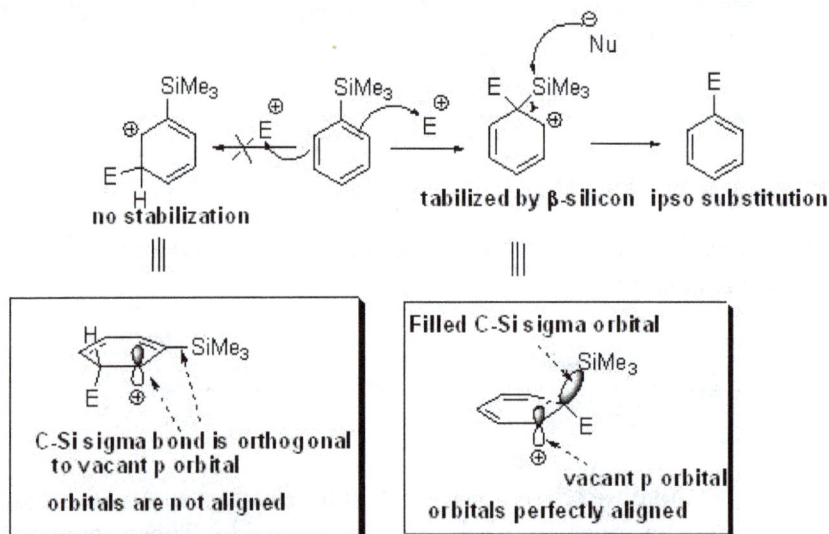

no stabilization   tabilized by β-silicon ipso substitution

C-Si sigma bond is orthogonal to vacant p orbital

orbitals are not aligned

Filled C-Si sigma orbital

vacant p orbital

orbitals perfectly aligned

An example for ipso substitution follows:

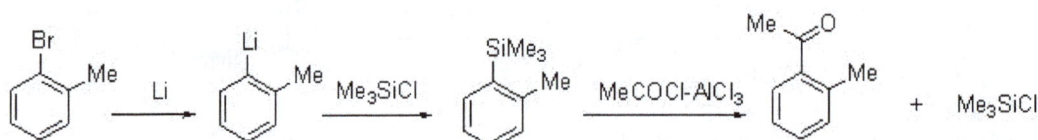

Allyl Silanes

Allylsilanes can be readily prepared from allyl halide via Grignard reaction.

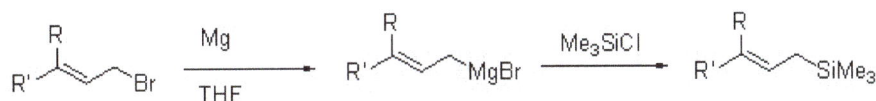

Allyl silanes are more reactive compared to vinyl silanes. This is because vinyl silanes have C-Si bonds orthogonal to the p orbitals of the alkene, in contrast, allyl silanes have C-Si bonds that can be parallel to the p orbitals of the double bond so that interaction can be possible. However, both react with electrophiles at the ipso atom occupied by silicon. In both cases a b -silylcation is an intermediate.

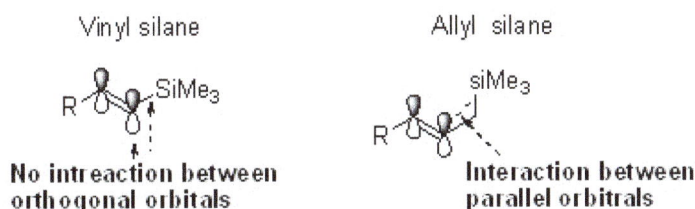

For example,

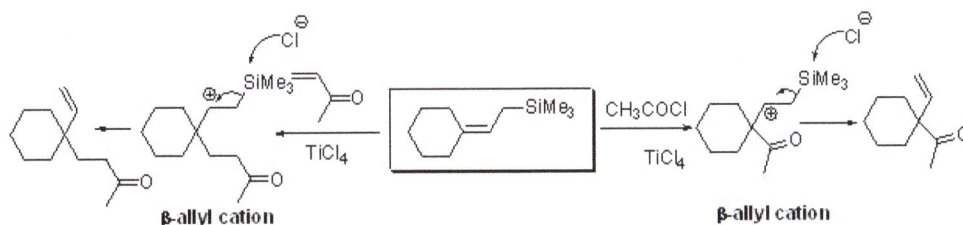

In optically active compounds, one enantiomer of the allyl silane gives one enantiomer of the product. The stereogenic centre next to silicon disappears and new one appears.

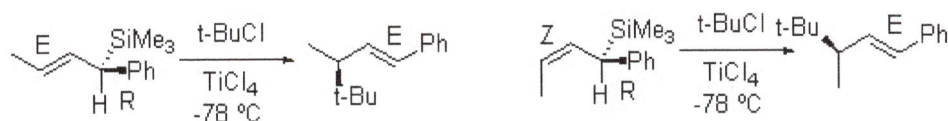

Allyl silanes also attack carbonyl compounds in the presence of Lewis acid which activates the carbonyl group.

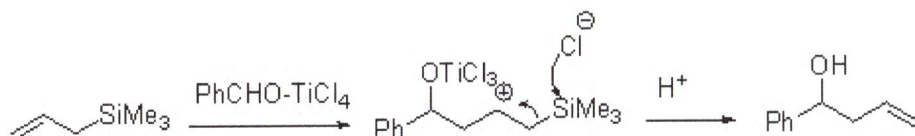

Silyl Epoxides

Silyl epoxides can be prepared from vinyl silanes with peroxy acids or from ketones.

Peterson Olefination Reaction

Peterson olefination, which is closely related to the Wittig reaction, can be carried out in two ways using either acid or base and the geometry of the alkenes can be controlled accordingly.

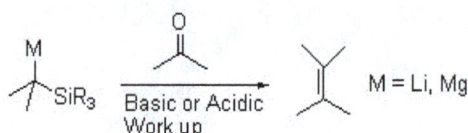

Mechanism

The reactions are anti under acidic conditions and syn under basic conditions. The stereoselectivity is due to involvement of a cyclic transition state under basic conditions, whereas under acidic conditions an acylic mechanism involves.

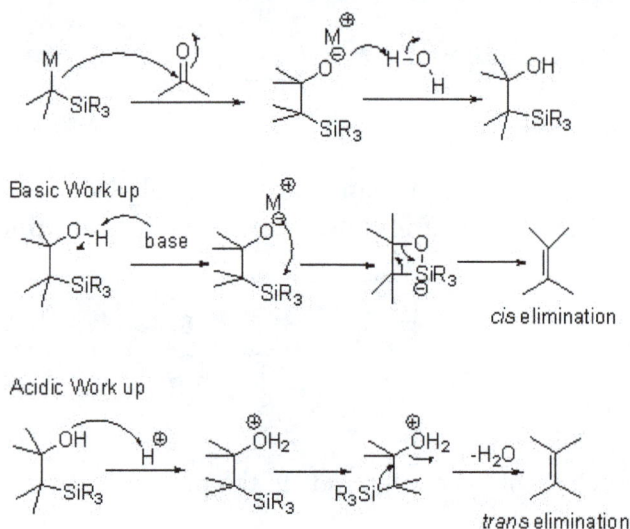

Examples

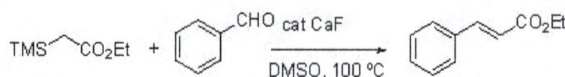

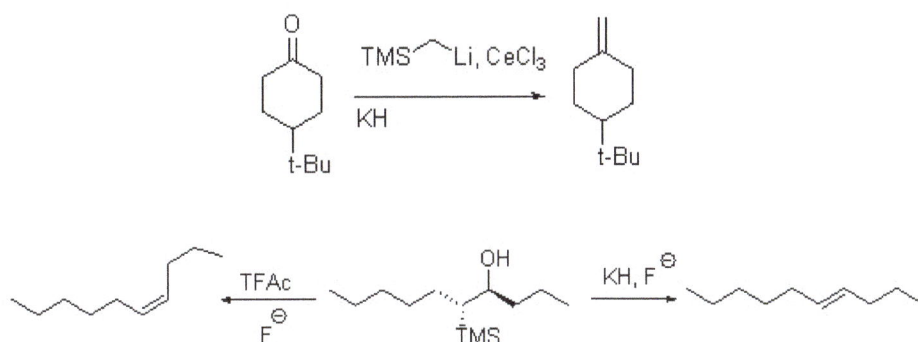

Boron Containing Compounds

Boron containing compounds find wide applications in organic synthesis. Borane (BH_3) is commercially available in the form of complexes generally with THF, Et_2O and Me_2S. It can also be prepared in situ by the reaction of $NaBH_4$ with $BF_3 \cdot OEt_2$ complex:

$$3\ NaBH_4\ +\ 4\ \overset{\ominus}{F_3}\overset{\oplus}{B}\text{-}OEt_2 \longrightarrow 3NaBH_4\ +\ 4\overset{\ominus}{H_3}\overset{\oplus}{B}\text{-}OEt_2$$

Organoboranes are synthesized by the addition reaction of borane to alkenes and alkynes.

Hydroboration of Alkenes

The reaction of borane with alkenes gives alkylboranes that readily proceed oxidation in the presence of alkaline hydrogen peroxide to yield alcohols. The conversion of C-B bond into a C-OH takes place with retention of stereochemistry.

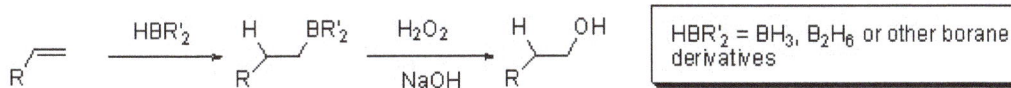

In case of sterically hindered alkenes such as trisubstituted ones, it is more difficult to add three alkenes to borane. This becomes the basis for the development of a variety of borane derivatives.

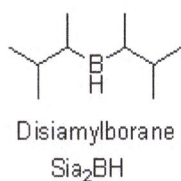

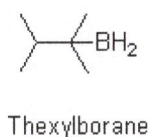

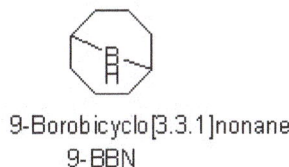

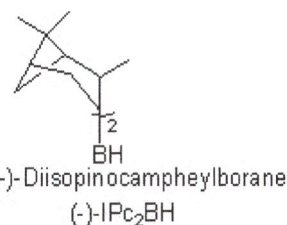

Disiamylborane
Sia_2BH

Thexylborane

9-Borobicyclo[3.3.1]nonane
9-BBN

(-)-Diisopinocampheylborane
(-)-IPc$_2$BH

Mechanism

The hydroboration reactions proceed by cis addition of hydrogen and boron to alkenes, probably via a four-centred cyclic transition state.

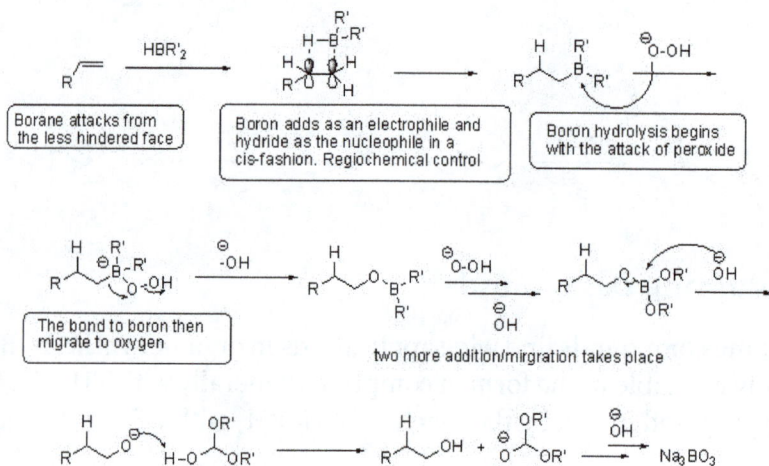

Borane attacks from the less hindered face

Boron adds as an electrophile and hydride as the nucleophile in a cis-fashion. Regiochemical control

Boron hydrolysis begins with the attack of peroxide

The bond to boron then migrate to oxygen

two more addition/mirgration takes place

Examples

BHCy$_2$, THF
NaBO$_3$ 4H$_2$O
H$_2$O

9-BBN, THF
H$_2$O$_2$, NaOH
EtOH

55%   15%

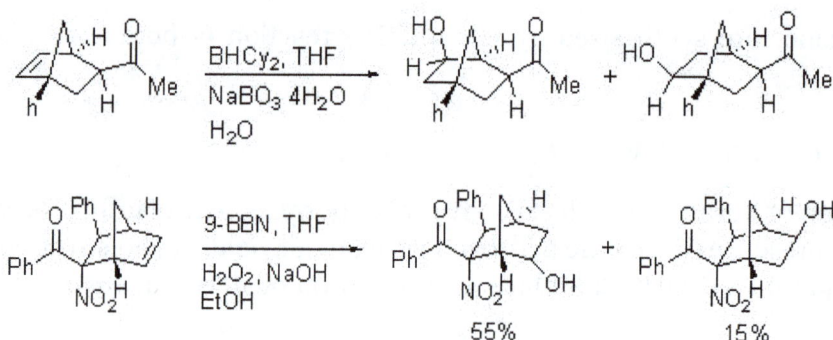

Reactions of Alkylboranes

Oxidation to Alcohols

Hydroboration and oxidative strategy provides an effective route for the transformation of alkenes to alcohols with retention of configuration at the boron-bearing carbon.

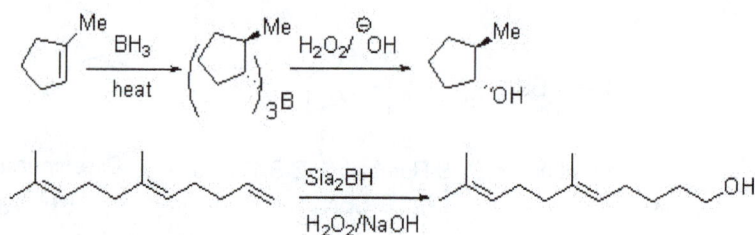

BH$_3$
heat

H$_2$O$_2$/$^{\ominus}$OH

Sia$_2$BH
H$_2$O$_2$/NaOH

Asymmetric version of this process has made a remarkable progress using chiral boranes with excellent enantioselectivity.

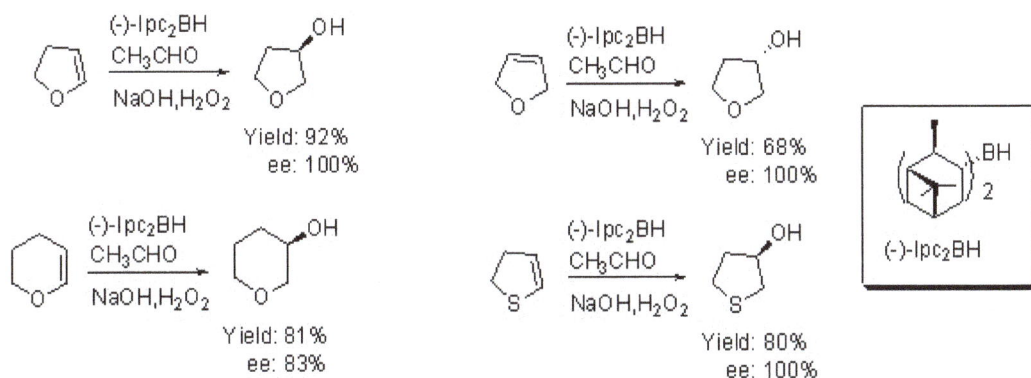

Yield: 92%
ee: 100%

Yield: 68%
ee: 100%

Yield: 81%
ee: 83%

Yield: 80%
ee: 100%

(-)-Ipc$_2$BH

Alkylboranes also undergo isomerization on heating to give products that contain the boron atom at the least hindered position of the alkyl chain.

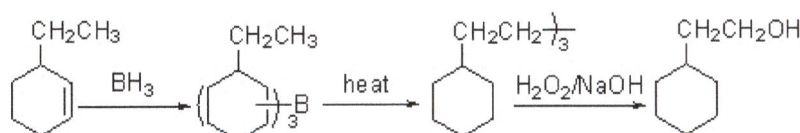

Coupling Reaction

Alkylboranes can be coupled using basic silver nitrate via an alkyl silver intermediate that affords a useful tool for the carbon-carbon bond formation.

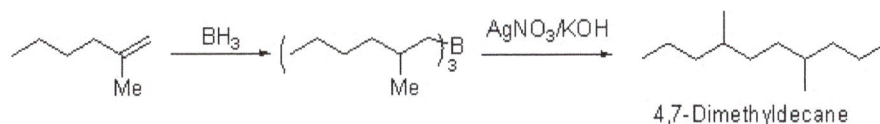

4,7-Dimethyldecane

Carbonylation: Formation of Alcohols, Aldehydes and Ketones

The reactions of organoboranes with carbon monoxide open up a variety of synthetic pathways for the preparation of alcohols, aldehydes and ketones. For example, 1,5,9-cyclododecatriene with B$_2$H$_6$ provides tricyclic borane that can be converted into tricylic alcohol via carbonylation and oxidation.

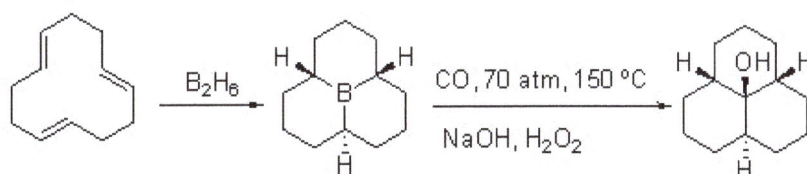

Mechanism

The reaction involves migration of alkyl groups from boron to the carbon atom of CO.

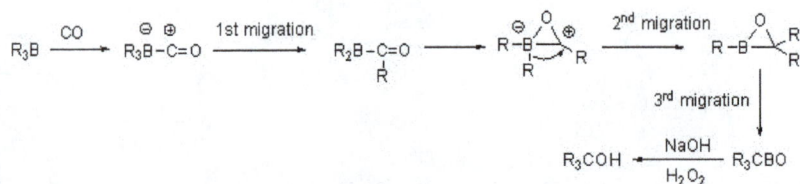

In presence of a small amount of water, migration of third alkyl group from boron to carbon can be inhibited to give dialkylketone.

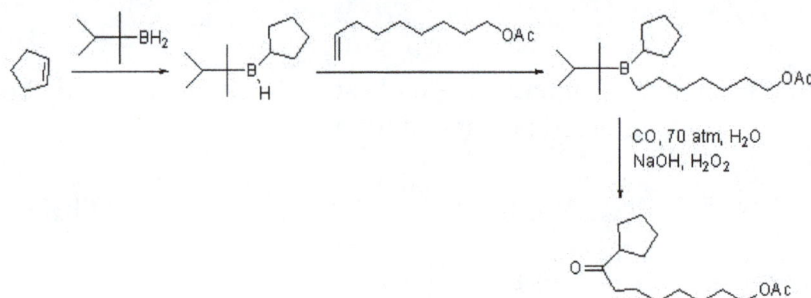

The carbonylation sequence can be modified to give aldehydes and primary alcohols.

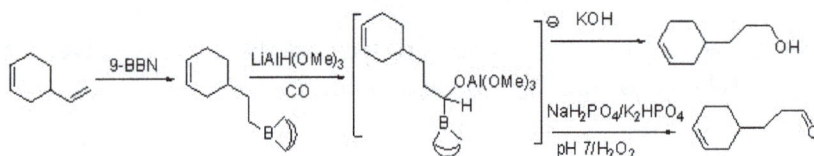

Conjugate Addition

Alkylboranes proceed conjugate addition with α,β -unsaturated aldehydes and ketones. The alkyl group of the borane undergoes 1,4-addition and boron is transferred to the oxygen, providing a boron enolate that on hydrolysis yields the product.

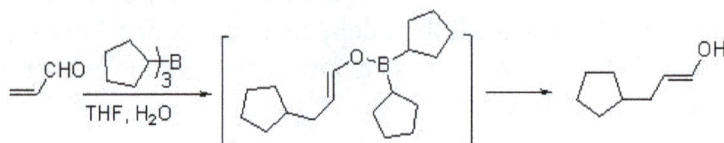

Allylation to Carbonyl Compounds

The reaction of allylboranes with carbonyl compounds has been well explored. Several studies have focused on asymmetric version of the process with excellent enantioselec-

tivity. For example, the reaction of (-)- β -allyl(diisopinocampheyl)borane with propio-
laldehyde gives hex-5-ene-1-yn-3-ol with 90% ee after the oxidation.

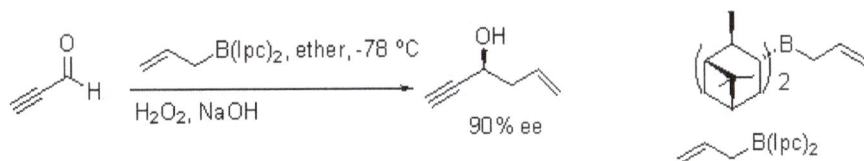

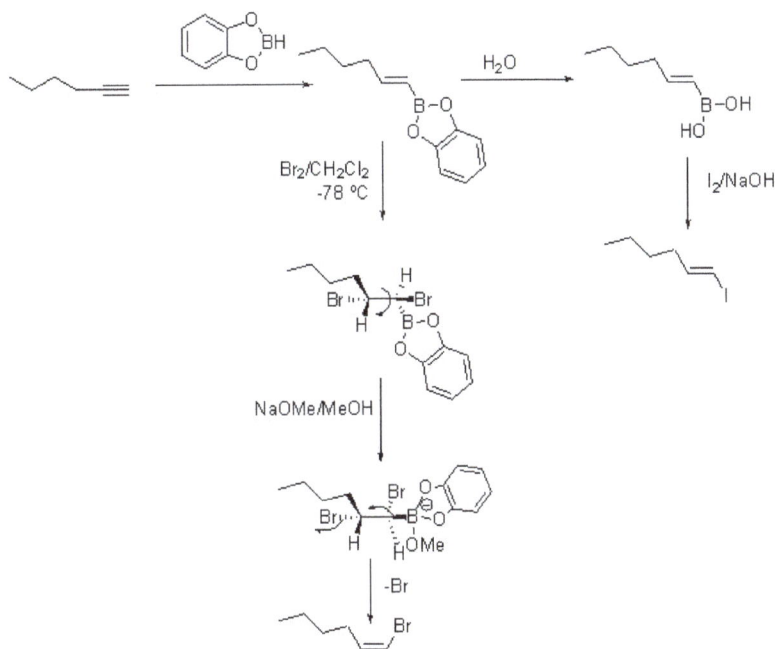

Vinylboranes

Borane with alkynes gives vinylboranes that serve as useful intermediates in organic synthesis. For example, 1-hexyne with catcholborane yields trans -1-alkenylborone
that can be converted into trans -vinyl iodide and cis -vinyl bromide that are substrate
precursors for the C-C coupling reactions using palladium catalysis. Vinylboranes can
also be converted into aldehydes, ketones or alkenes and the reactivity and selectivity
depend on the nature of organoboranes used.

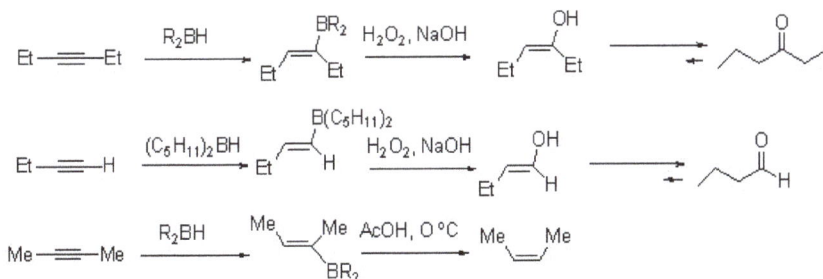

Suzuki Coupling

The coupling of organic boronic acid with halides or triflates using palladium-catalysis leads to a powerful protocol for the carbon-carbon bond formation.

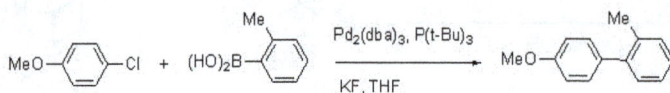

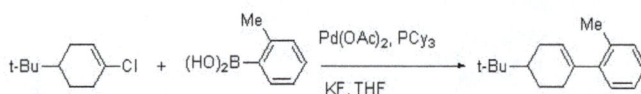

Examples:

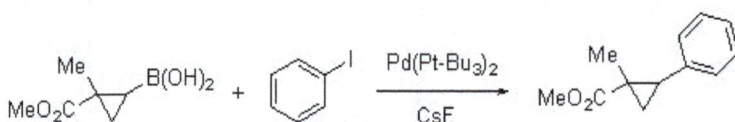

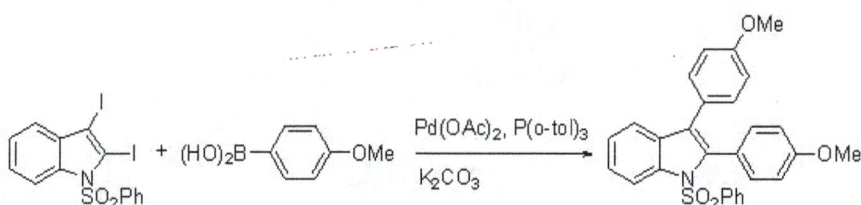

Organotin Compounds

Tin-Containing Compounds

The synthesis of organotin compounds is similar to that of organosilicones. The reaction of Grignard reagent with bis(tributyltin)oxide gives alkyl tributyltin. The polarity can be reversed and stannyl lithium can add to organic electrophiles.

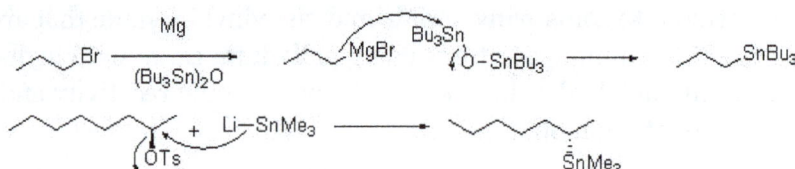

The hydrostannylation of an alkyne with tin hydride affords kinetically controlled Z-vinyl stannane. If there is an excess of tin hydride or sufficient radicals are present, isomerization may take place to afford the more stable E -isomer.

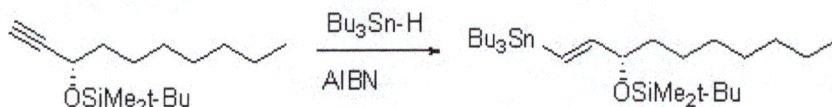

Mechanism

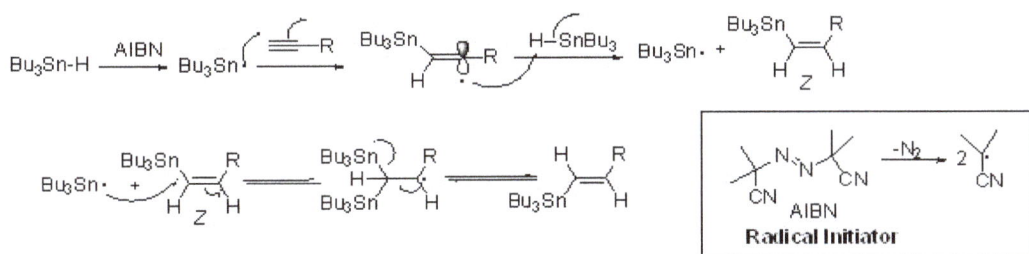

Stille Coupling

Palladium-catalyzed cross-coupling of vinyl stannanes with vinyl halides or triflates give dienes. The reaction functions under relatively neutral conditions and compatible with many functional groups. Both inter- and intramolecular versions of the reactions have been explored and find extensive applications in natural product synthesis.

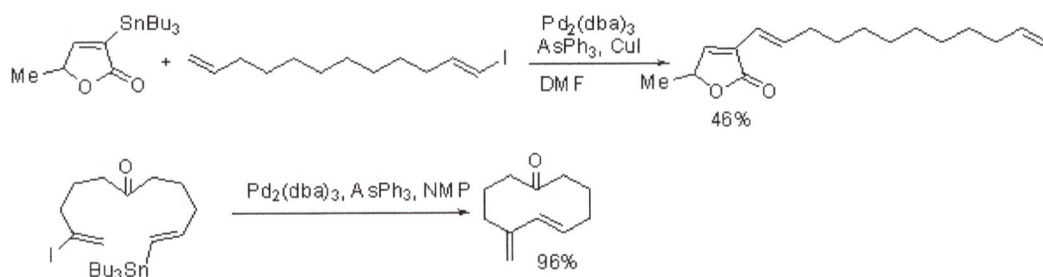

Mechanism

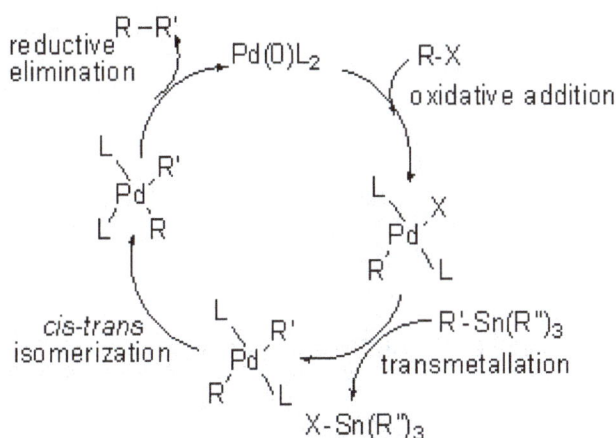

If the electrophile is a vinyl triflate, the addition of LiCl to the reaction is essential because the chloride may displace triflate form the palladium σ-complex. The transmetallation takes place with chloride on palladium and not with triflate.

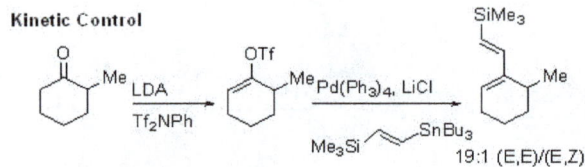

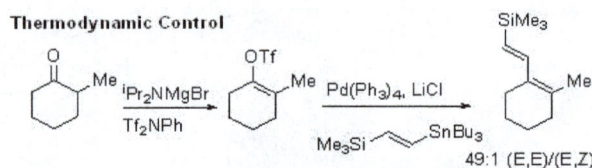

Kinetic Control

19:1 (E,E)/(E,Z)

Thermodynamic Control

49:1 (E,E)/(E,Z)

Examples

95%

38%

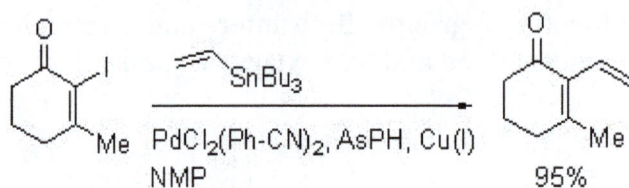

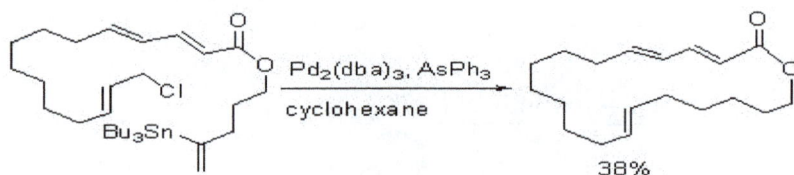

Reactions of Allyl Stannanes

Allyl stannanes are important reagents because they can be used for allylation to aldehydes with excellent stereocontrol.

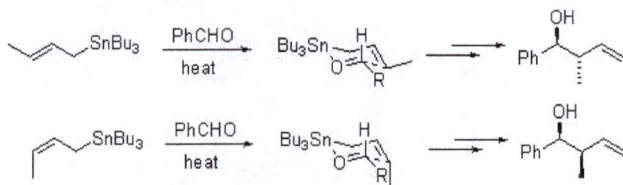

The asymmetric allyllation of aldehydes with ally stannanes has also been explored.

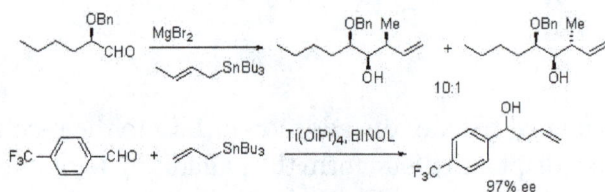

10:1

97% ee

Reactions of Tributyltinhydride

Tributyltinhydride (Bu$_4$SnH) is a versatile reagent for the removal of halogen (I and Br) from alkyl halides. The reaction is carried out either in the presence of light or AIBN which is a radical initiator.

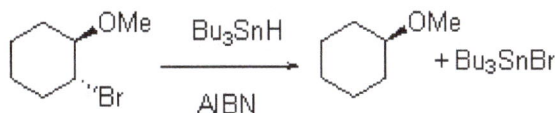

Mechanism

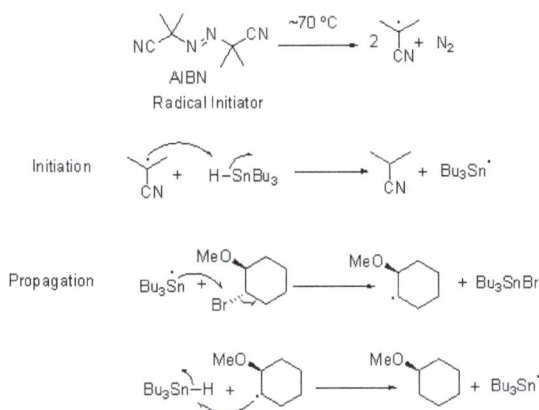

The alkyl radical can also undergo addition to alkenes to give alkane via the formation new carbon-carbon bond. Both inter- and intramolecular versions of this process have been explored. For example, Bu$_3$SnH mediated addition of nucleophilic radical to electrophilic alkene can be accomplished in good yield.

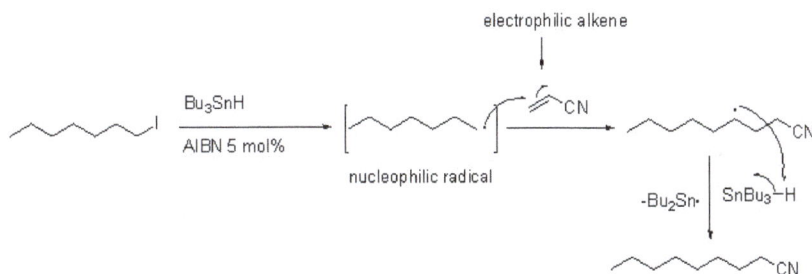

Similarly, the Bu$_3$SnH mediated addition of electrophilic radical to nucleophilic alkene can also be accomplished.

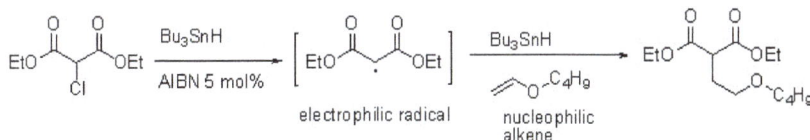

Examples

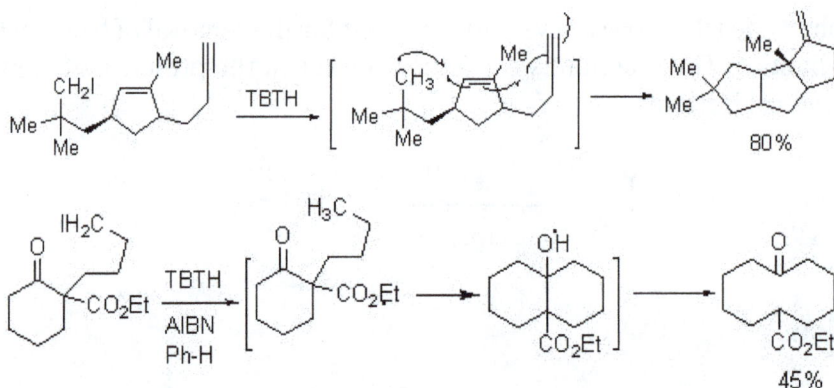

References

- Carey, J.S.; Laffan, D.; Thomson, C. & Williams, M.T. (2006). "Analysis of the reactions used for the preparation of drug candidate molecules". Org. Biomol. Chem. 4: 2337–2347. PMID 16763676. doi:10.1039/B602413K

- Woodward, R. B.; Cava, M. P.; Ollis, W. D.; Hunger, A.; Daeniker, H. U.; Schenker, K. (1954). "The Total Synthesis of Strychnine". Journal of the American Chemical Society. 76 (18): 4749–4751. doi:10.1021/ja01647a088

- Arnoldischen Buchhandlung, Dresden and Leipzig, 1827. ISBN 1-148-99953-1. Brief English commentary in English can be found in Bent Soren Jorgensen "More on Berzelius and the vital force" J. Chem. Educ., 1965, vol. 42, p 394. doi:10.1021/ed042p394

- Todd, Matthew H. (2005). "Computer-aided Organic Synthesis". Chemical Society Reviews. 34 (3): 247–266. PMID 15726161. doi:10.1039/b104620a

- Nguyen, Lien Ai; He, Hua; Pham-Huy, Chuong (2016-11-20). "Chiral Drugs: An Overview". International Journal of Biomedical Science : IJBS. 2 (2): 85–100. ISSN 1550-9702. PMC 3614593. PMID 23674971

- Service. R.F. (2001). "Science Awards Pack a Full House of Winners" (print, online science news). Science. 294 (5542; October 19): 503–505. PMID 11641480. doi:10.1126/science.294.5542.503b. Retrieved 2 March 2016

- Noyori, R.; Ikeda, T.; Ohkuma, T.; Widhalm, M.; Kitamura, M.; Takaya, H.; Akutagawa, S.; Sayo, N.; Saito, T. "Stereoselective hydrogenation via dynamic kinetic resolution". Journal of the American Chemical Society. 111 (25): 9134–9135. doi:10.1021/ja00207a038

- Smith, Cory. "Petrochemicals". American Fuel & Petrochemical Manufacturers. American Fuel & Petrochemical Manufacturers. Retrieved 18 December 2016

- Spencer L. Seager, Michael R. Slabaugh. Chemistry for Today: general, organic, and biochemistry. Thomson Brooks/Cole, 2004, p. 342. ISBN 0-534-39969-X

- Knowles, William S. (2002-06-17). "Asymmetric Hydrogenations (Nobel Lecture)". Angewandte Chemie International Edition. 41 (12): 1998–2007. ISSN 1521-3773. doi:10.1002/1521-3773(20020617)41:123.0.CO;2-8

- Gao, Yun; Klunder, Janice M.; Hanson, Robert M.; Masamune, Hiroko; Ko, Soo Y.; Sharpless, K. Barry (1987-09-01). "Catalytic asymmetric epoxidation and kinetic resolution: modified procedures including in situ derivatization". Journal of the American Chemical Society. 109 (19): 5765–5780. ISSN 0002-7863. doi:10.1021/ja00253a032

- "IUPAC Blue Book, Table 28(a) Carboxylic acids and related groups. Unsubstituted parent struc-tures". Retrieved 2009-11-22

- Newman, D. K.; Banfield, J. F. (2002). "Geomicrobiology: How Molecular-Scale Interactions Underpin Biogeochemical Systems". Science. 296 (5570): 1071–1077. PMID 12004119. doi:10.1126/science.1010716

Synthesis of Naturally Occurring Compounds

Naturally occurring compounds are helpful in the development of synthetic targets that can initialize a biological process in the host. Pharmaceuticals are an important sector of the application of organic synthesis of compounds. The organic structure of a drug is important to understand its effects on a patient. This chapter has been carefully written to provide an easy understanding of the varied facets of naturally occurring compounds.

Reserpine

Reserpine (also known by trade names Raudixin, Serpalan, Serpasil) is an indole alkaloid, antipsychotic, and antihypertensive drug that has been used for the control of high blood pressure and for the relief of psychotic symptoms, although because of the development of better drugs for these purposes and because of its numerous side-effects, it is rarely used today. The antihypertensive actions of reserpine are a result of its ability to deplete catecholamines (among other monoamine neurotransmitters) from peripheral sympathetic nerve endings. These substances are normally involved in controlling heart rate, force of cardiac contraction and peripheral vascular resistance.

Reserpine-mediated depletion of monoamine neurotransmitters in the synapses is often cited as evidence to the theory that depletion of the monoamine neurotransmitters causes subsequent depression in humans (c.f. monoamine hypothesis). However, this claim is not without controversy. The reserpine-induced depression is considered by some researchers to be a myth, while others claim that teas made out of the plant roots containing reserpine have a calming, sedative action that can actually be considered *anti*depressant. Notably, reserpine was the first compound shown to be an effective antidepressant in a randomized placebo-controlled trial.

Moreover, reserpine has a peripheral action in many parts of the body, resulting in a preponderance of the effects of the cholinergic part of the autonomous nervous system on the GI tract, smooth muscles, blood vessels, etc.

Mechanism of Action

Reserpine irreversibly blocks the vesicular monoamine transporter (VMAT). This normally transports free intracellular norepinephrine, serotonin, and dopamine in the presynaptic nerve terminal into presynaptic vesicles for subsequent release into the synaptic cleft ("exocytosis"). Unprotected neurotransmitters are metabolized by MAO (as well as by COMT) in the cytoplasm and consequently never excite the post-synaptic cell.

It may take the body days to weeks to replenish the depleted VMAT, so reserpine's effects are long-lasting.

This depletion of dopamine can lead to drug-induced parkinsonism.

Biosynthetic pathway

Tryptophan is the starting material in the biosynthetic pathway of reserpine, and is converted to tryptamine by tryptophan decarboxylase enzyme. Tryptamine is combined with secologanin in the presence of strictosidine synthetase enzyme and yields strictosidine. Various enzymatic conversion reactions lead to the synthesis of reserpine from strictosidine.

History

Reserpine was isolated in 1952 from the dried root of *Rauwolfia serpentina* (Indian snakeroot), which had been known as *Sarpagandha* and had been used for centuries in India for the treatment of insanity, as well as fever and snakebites — Mahatma Gandhi used it as a tranquilizer. It was first used in the United States by Robert Wallace Wilkins in 1950. Its molecular structure was elucidated in 1953 and natural configuration published in 1955. It was introduced in 1954, two years after chlorpromazine. The first total synthesis was accomplished by R. B. Woodward in 1958.

Reserpine almost irreversibly blocks the uptake (and storage) of norepinephrine (i.e. noradrenaline) and dopamine into synaptic vesicles by inhibiting the Vesicular Monoamine Transporters (VMAT).

Reserpine has been discontinued in the UK for some years due to its numerous interactions and side effects.

Reserpine was also highly influential in promoting the thought of a biogenic amine hypothesis of depression .

Uses Today

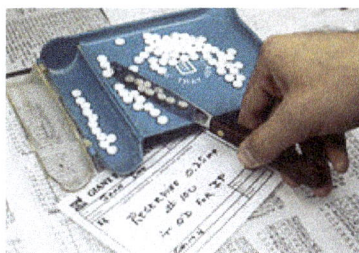

Seen is the drug Reserpine in tablet form being counted out to fill a prescription in a pharmacy.

Medical

Reserpine is one of the few antihypertensive medications that have been shown in randomized controlled trials to reduce mortality: The Hypertension Detection and Follow-up Program, the Veterans Administration Cooperative Study Group in Anti-hypertensive Agents, and the Systolic Hypertension in the Elderly Program.

Reserpine is rarely used in the management of hypertension today. Reserpine is listed as an option by the JNC 7. Reserpine is a second-line adjunct agent for patients who are uncontrolled on a diuretic when cost is an issue.

It is also used to treat symptoms of dyskinesia in patients suffering from Huntington's disease.

In some countries reserpine is still available as part of combination drugs for the treatment of hypertension, in most cases they contain also a diuretic and/or a vasodilator like hydralazine. These combinations are currently regarded as second choice drugs. The daily dose of reserpine in antihypertensive treatment is as low as 0.1 to 0.25 mg. The use of reserpine as an antipsychotic drug had been nearly completely abandoned, but more recently it made a comeback as adjunctive treatment, in combination with other antipsychotics, so that more refractory patients get dopamine blockade from the other antipsychotic, and dopamine depletion from reserpine. Doses for this kind of adjunctive goal can be kept low, resulting in better tolerability. Originally, doses of 0.5 mg to 40 mg daily were used to treat psychotic diseases. Doses in excess of 3 mg daily often required use of an anticholinergic drug to combat excessive cholinergic activity in many parts of the body as well as parkinsonism. For adjunctive treatment, doses are typically kept at or below 0.25 mg twice a day.

Veterinary and Other

Reserpine may be used as a sedative for horses.

Another frequent use of reserpine is in the field of mass spectrometry where it is widely used as a reference standard owing to its availability, ease of ionization under electrospray conditions and stability in solution.

Reserpine is used as a long-acting tranquilizer to subdue excitable or difficult horses and has been used illicitly for the sedation of show horses, for-sale horses, and in other circumstances where a "quieter" horse might be desired.

Reserpine is no longer available in the United Kingdom however it is still available in the United States

Side Effects

At doses of less than 0.2 mg/day, reserpine has few side effects, the most common of which is nasal congestion.

There has been much concern about reserpine causing depression leading to suicide. However, this was reported in uncontrolled studies using doses averaging 0.5 mg per day.

Reserpine can cause: nasal congestion, nausea, vomiting, weight gain, gastric intolerance, gastric ulceration (due to increased cholinergic activity in gastric tissue and impaired mucosal quality), stomach cramps and diarrhea are noted. The drug causes hypotension and bradycardia and may worsen asthma. Congested nose and erectile dysfunction are other consequences of alpha-blockade. Depression can occur at any dose and may be severe enough to lead to suicide. Other central effects are a high incidence of drowsiness, dizziness, and nightmares. Parkinsonism occurs in a dose dependent manner. General weakness or fatigue is quite often encountered. High dose studies in rodents found reserpine to cause fibroadenoma of the breast and malignant tumors of the seminal vesicles among others. Early suggestions that reserpine causes breast cancer in women (risk approximately doubled) were not confirmed. It may also cause hyperprolactinemia.

Reserpine passes into breast milk and is harmful to breast-fed infants, and should therefore be avoided during breastfeeding if possible.

It also produces a dangerous decline in blood pressure at doses needed for treatment.

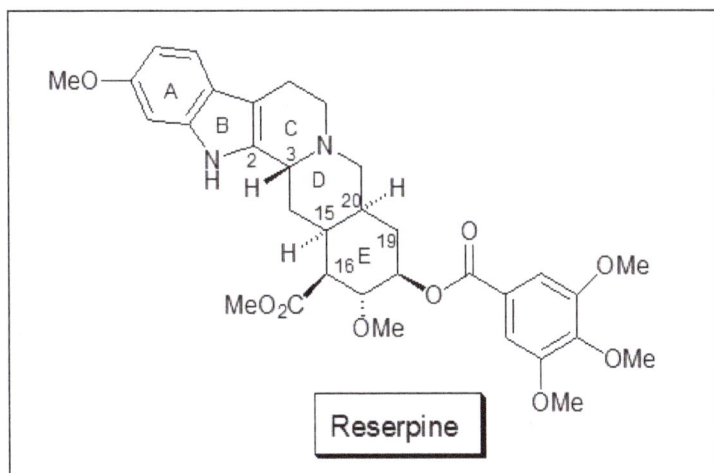

Reserpine

It is found, with other alkaloids, in the roots of the plant genus Rauwolfia and used in the treatment of some mental disorders as well as for the reduction of hypertension.

Retrosynthetic Analysis

Retrosynthetic analysis is a technique for solving problems in the planning of organic syntheses. This is achieved by transforming a target molecule into simpler precursor structures without assumptions regarding starting materials. Each precursor material is examined using the same method. This procedure is repeated until simple or commercially available structures are reached. E.J. Corey formalized this concept in his book *The Logic of Chemical Synthesis.*

The power of retrosynthetic analysis becomes evident in the design of a synthesis. The goal of retrosynthetic analysis is structural simplification. Often, a synthesis will have more than one possible synthetic route. Retrosynthesis is well suited for discovering different synthetic routes and comparing them in a logical and straightfoward fashion. A database may be consulted at each stage of the analysis, to determine whether a component already exists in the literature. In that case, no further exploration of that compound would be required.

Definitions

Disconnection

A retrosynthetic step involving the breaking of a bond to form two (or more) synthons.

Retron

A minimal molecular substructure that enables certain transformations.

Retrosynthetic tree

A directed acyclic graph of several (or all) possible retrosyntheses of a single target.

Synthon

An idealized molecular fragment. A synthon and the corresponding commercially available synthetic equivalent are shown below:

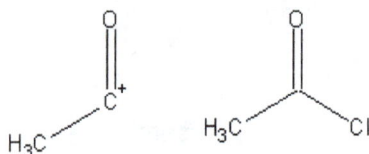

Target

   The desired final compound.

Transform

   The reverse of a synthetic reaction; the formation of starting materials from a single product.

Example

An example will allow the concept of retrosynthetic analysis to be easily understood.

Bond "breaks" here

Synthons

Synthetic equivalents

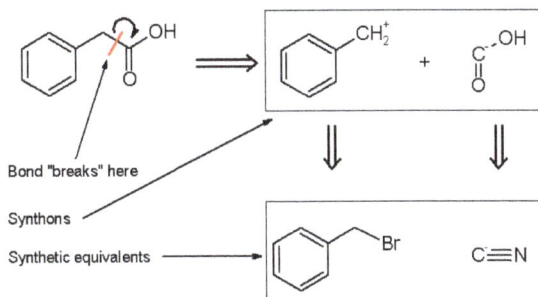

In planning the synthesis of phenylacetic acid, two synthons are identified. A nucleophilic "-COOH" group, and an electrophilic "PhCH$_2$⁺" group. Of course, both synthons do not exist per se; synthetic equivalents corresponding to the synthons are reacted to produce the desired product. In this case, the cyanide anion is the synthetic equivalent for the ⁻COOH synthon, while benzyl bromide is the synthetic equivalent for the benzyl synthon.

The synthesis of phenylacetic acid determined by retrosynthetic analysis is thus:

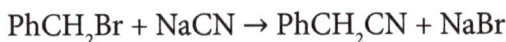

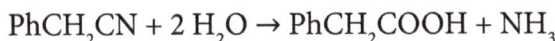

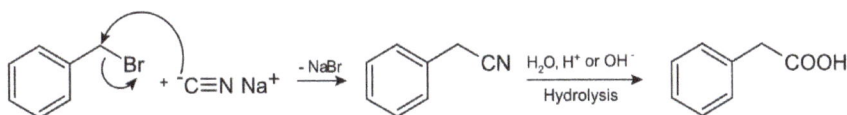

$$PhCH_2Br + NaCN \rightarrow PhCH_2CN + NaBr$$

$$PhCH_2CN + 2 H_2O \rightarrow PhCH_2COOH + NH_3$$

In fact, phenylacetic acid has been synthesized from benzyl cyanide, itself prepared by the analogous reaction of benzyl chloride with sodium cyanide.

Strategies

Functional Group Strategies

Manipulation of functional groups can lead to significant reductions in molecular complexity.

Stereochemical Strategies

Numerous chemical targets have distinct stereochemical demands. Stereochemical transformations (such as the Claisen rearrangement and Mitsunobu reaction) can remove or transfer the desired chirality thus simplifying the target.

Structure-goal Strategies

Directing a synthesis toward a desirable intermediate can greatly narrow the focus of an analysis. This allows bidirectional search techniques.

Transform-based Strategies

The application of transformations to retrosynthetic analysis can lead to powerful reductions in molecular complexity. Unfortunately, powerful transform-based retrons are rarely present in complex molecules, and additional synthetic steps are often needed to establish their presence.

Topological Strategies

The identification one or more key bond disconnections may lead to the identification of key substructures or difficult to identify rearrangement transformations in order to identify the key structures .

- Disconnections that preserve ring structures are encouraged.

- Disconnections that create rings larger than 7 members are discouraged.

Retrosynthetic Analysis

The strategy was based on building five contiguous stereocentres into a decalin derivative that could be opened to a monocyclic compound to form ring E.

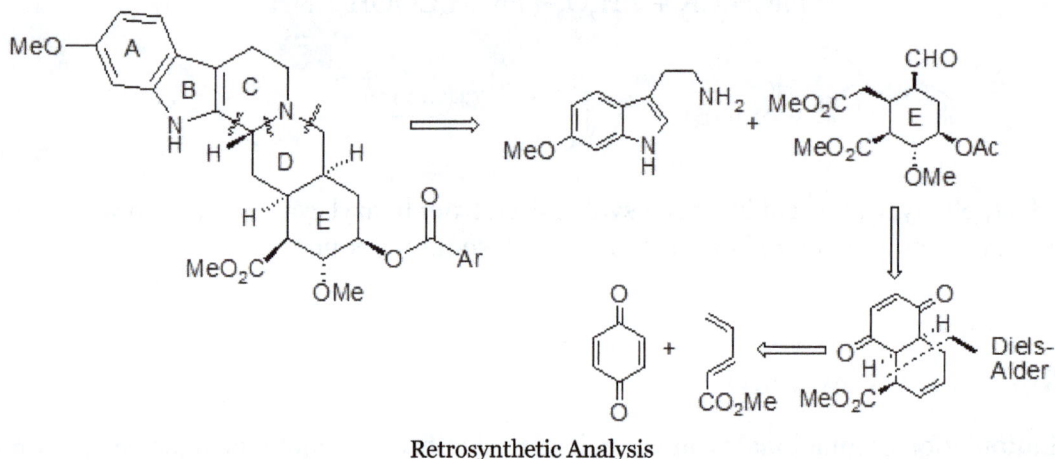

Retrosynthetic Analysis

Total Synthesis

- The Diels-Alder reaction can lead to the ring junction having cis stereochemistry and the carboxyl group lie on the same side as the rings with respect to the ring junction (i) . This step fixes the stereochemistry at C 15, C 16 and C20 of reserpine.

- NaBH4 reduction of the less hindered of the two carbonyl groups of 2 can provide 3 (ii). The epoxidation of the isolated double bond with mCPBA at the less hindered side can afford 4 (iii) that could undergo dehydration to give the lactone 5 (iv).

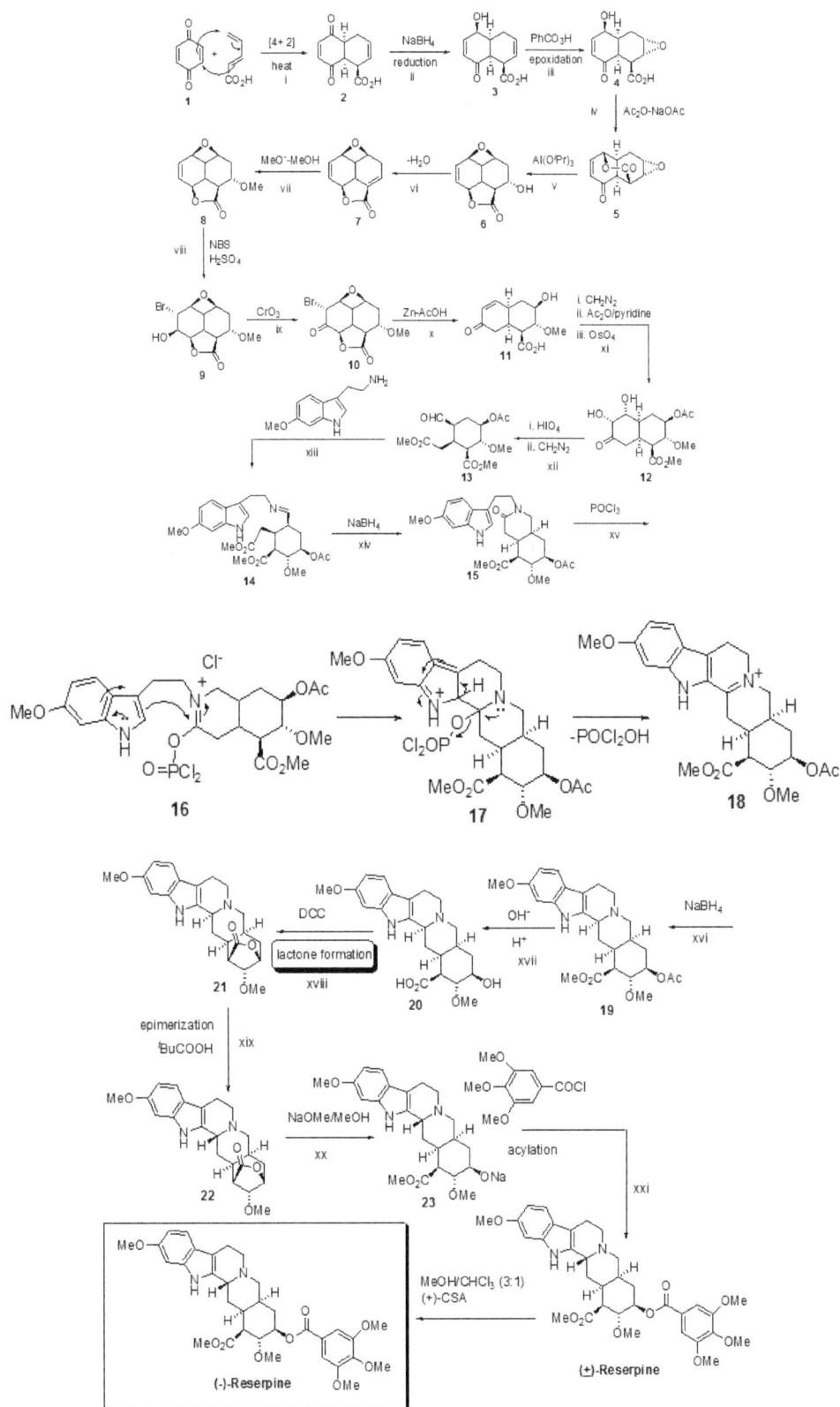

Total Synthesis

- Meerwein-Ponndorf-Verley reduction of 5 could convert the keto group into hydroxyl that can displace on the carbonyl of the six membered lactone ring, giving a five membered lactone, and the hydroxyl group so released can open the epoxide ring to afford 6.

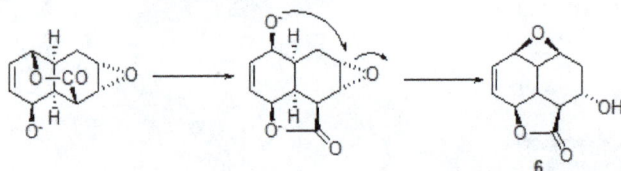

- Dehydration of 6 can give α,β -unsaturated carbonyl compound 7 that could undergo conjugate addition at the less hindered α -side with methoxide to give 8 (vi and vii).

- NBS in acid could approach α -side of 8 to give a brominium ion that could be opened by water to give the biaxial bromo-alcohol 9 (viii) that could undergo mild oxidation to afford 10 (ix).

- Zn in AcOH can bring the reductive opening of both the lactone and the strained ether of 10 to give 11 (x).

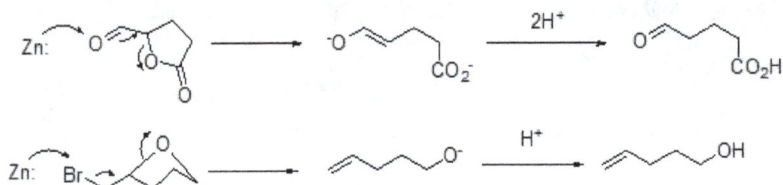

- Esterification of the carboxyl group using diazomethane,acetylation of the alcohol group using Ac_2O and dihydroxylation of the double bond can give 12 (xi) that could undergo oxidative cleavage followed by esterification of the new carboxyl group with diazomethane to give 13 (xii).

- Schiff base formation of 13 with 6-methoxytryptamine can give 14 that could be converted into 15 by $NaBH_4$ reduction of the imine double bond (xiii and xiv). Treatment of 15 with $POCl_3$ can bring ring closure as in the Bischler-Napieralski synthesis of isoquinoline, providing an imminium salt 18 via 16 and 17 (xv), which could be reduced using $NaBH_4$ to give 19 (xvi).

- Base hydrolysis of 19 can give 20 having free OH and COOH groups that could be joined to give a lactone 21 using DCC (xvii and xviii). Epimerization of the less stable 21 using t-butyric acid can give the required more stable 22 that could be converted into (±)-reserpine by opening of the lactone with MeOH followed by acylation using 3,4,5-trimethoxybenzoyl chloride. The (±)-reserpine could be resolved using CSA in a 3:1 mixture of MeOH and $CHCl_3$.

Phenoxymethylpenicillin

Phenoxymethylpenicillin, also known as penicillin V, is an antibiotic useful for the treatment of a number of bacterial infections. Specifically it is used for the treatment of strep throat, otitis media, and cellulitis. It is also used to prevent rheumatic fever and to prevent infections following removal of the spleen. It is given by mouth.

Side effects include diarrhea, nausea, and allergic reactions including anaphylaxis. It is not recommended in those with a history of penicillin allergy. It is relatively safe for use during pregnancy. It is in the penicillin and beta lactam family of medications. It usually results in bacterial death.

Phenoxymethylpenicillin was first made in 1948. It is on the World Health Organization's List of Essential Medicines, the most effective and safe medicines needed in a health system. It is available as a generic medication. The wholesale cost in the developing world is about 0.05 to 0.96 USD per day. In the United States a course of treatment costs less than 25 USD.

Medical Uses

Specific indications for phenoxymethylpenicillin include:

- Infections caused by *Streptococcus pyogenes*

  o Tonsillitis

  o Pharyngitis

  o Skin infections

- Anthrax (mild uncomplicated infections)

- Lyme disease (early stage in pregnant women or young children)

- Rheumatic fever (primary and secondary prophylaxis)

- Streptococcal skin infections

- Spleen disorders (pneumococcal infection prophylaxis)

- Initial treatment for Dental Abscesses

- Moderate-to-severe gingivitis (with metronidazole)

- Avulsion injuries of teeth (as an alternative to tetracycline)

- Blood infection prophylaxis in children with sickle cell disease.

Penicillin V is sometimes used in the treatment of odontogenic infections.

It is less active than benzylpenicillin (penicillin G) against Gram-negative bacteria. Phenoxymethylpenicillin has a range of antimicrobial activity against Gram-positive bacteria that is similar to that of benzylpenicillin and a similar mode of action, but it is substantially less active than benzylpenicillin against Gram-negative bacteria.

Phenoxymethylpenicillin is more acid-stable than benzylpenicillin, which allows it to be given orally.

Phenoxymethylpenicillin is usually used only for the treatment of mild to moderate infections, and not for severe or deep-seated infections since absorption can be unpredictable. Except for the treatment or prevention of infection with *Streptococcus pyogenes* (which is uniformly sensitive to penicillin), therapy should be guided by bacteriological studies (including sensitivity tests) and by clinical response. People treated initially with parenteral benzylpenicillin may continue treatment with phenoxymethylpenicillin by mouth once a satisfactory response has been obtained.

It is not active against beta-lactamase-producing bacteria, which include many strains of *Staphylococci*.

Adverse Effects

Phenoxymethylpenicillin is usually well tolerated but may occasionally cause transient nausea, vomiting, epigastric distress, diarrhea, constipation, acidic smell to urine and black hairy tongue. A previous hypersensitivity reaction to *any* penicillin is a contraindication.

Mechanism of Action

It exerts a bactericidal action against penicillin-sensitive microorganisms during the stage of active multiplication. It acts by inhibiting the biosynthesis of cell-wall peptidoglycan.

Compendial Status

- British Pharmacopoeia

Total Synthesis

Total synthesis is the complete chemical synthesis of a complex molecule, often a natural product, from simple, commercially available precursors. It usually refers to a process not involving the aid of biological processes, which distinguishes it from semisynthesis. The target molecules can be natural products, medicinally important active ingredients, or organic compounds of theoretical interest. Often the aim is to discover new route of synthesis for a target molecule for which there already exist known routes. Sometimes no route exists and the chemist wishes to find a viable route for the first time. One important purpose of total synthesis is the discovery of new chemical reactions and new chemical reagents.

Scope and Definitions

The moniker of total synthesis is less frequently, but nevertheless accurately applied to the synthesis of natural polypeptides and polynucleotides; for instance, the peptide hormones oxytocin and vasopressin were isolated, and their total syntheses first reported, in 1954.

Aims

Although untrue from an historical perspective total synthesis in the modern are has largely been an academic endeavour (in terms of manpower applied to problems), although industrial concerns may pick up particular avenues of total synthesis efforts, and expend considerable resources on particular natural product targets, especially in cases where semi-synthesis can be applied to complex, natural product-derived drugs. Even so, there is continuing discussion regarding the value of total synthesis as an academic enterprise, some aspects of which are summarised here.

Total synthesis projects often require a variety of reactions, and so efforts to achieve complex total syntheses serve to prepare chemists for pursuits in pharmaceutical discovery chemistry, in particular, as well as in process chemistry, in both cases, where comprehensive knowledge of chemical reactions and a strong and accurate chemical intuition are important qualifications.

History

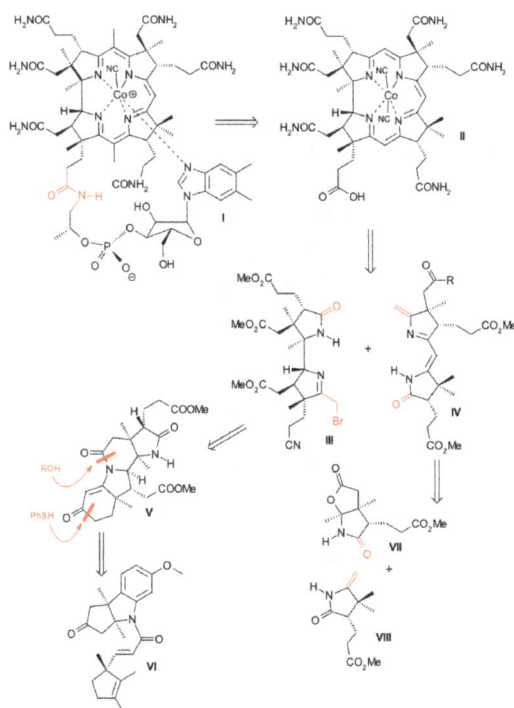

Vitamin B$_{12}$ total synthesis: Retrosynthetic analysis.

Analysis of the Woodward-Eschenmoser total synthesis that was reported in two variants, by these groups, in 1972. The work involved more than 100 PhD trainees and post-doctoral fellows, from 19 different nations. The retrosynthesis presents the disassembly of the target vitamin in a manner that makes chemical sense for its eventual forward construction. The target, Vitamin B_{12} (I), is envisioned being prepared by the simple addition of its tail, which had earlier been shown to be feasible. The needed precursor, cobyric acid (II), then becomes the target. This acid constitutes the "corrin core" of the vitamin, and its preparation was envisaged possible via two pieces, a "western" part copses of the A and D rings (III) and an "eastern" part composed of the B and C rings (*IV*). The restrosynthetic analysis then envisions the starting materials required to make these two complex parts, the yet complex molecules V–VIII.

Friedrich Wöhler discovered that an organic substance, urea, could be produced from inorganic starting materials in 1828. This was an important conceptual milestone in chemistry, as it was the first example of a synthesis of a substance known earlier, only as a byproduct of living processes. Wohler obtained urea by treating silver cyanate with ammonium chloride, a simple, one-step synthesis:

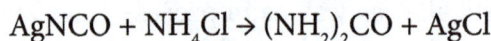

$$AgNCO + NH_4Cl \rightarrow (NH_2)_2CO + AgCl$$

Camphor was a scarce and expensive natural product with a worldwide demand. Haller and Blanc synthesized camphor from camphor acid; however, the precursor, camphoric acid, was of unknown structure. When Finnish chemist Gustav Komppa synthesized camphoric acid from diethyl oxalate and 3,3-dimethylpentanoic acid in 1904, the structure of these precursors allowed chemists at the time to infer the complicated ring structure of camphor. Shortly thereafter, William Perkin published another synthesis of camphor. The work on the total chemical synthesis of camphor allowed Komppa to begin industrial production of this compound, in Tainionkoski, Finland, in 1907.

The American chemist Robert Burns Woodward was a pre-eminent figure in developing total syntheses of complex organic molecules, where targets of his included cholesterol, cortisone, strychnine, lysergic acid, reserpine, chlorophyll, colchicine, vitamin B_{12} and prostaglandin F-2a.

Vincent du Vigneaud was awarded the 1955 Nobel Prize in Chemistry for the total synthesis of the natural polypeptide oxytocin and vasopressin, reported in 1954, with the citation: "for his work on biochemically important sulphur compounds, especially for the first synthesis of a polypeptide hormone."

Another gifted chemist is Elias James Corey who won the Nobel Prize in Chemistry in 1990 for lifetime achievement in total synthesis and the development of retrosynthetic analysis.

Examples

One classic in total synthesis is quinine total synthesis, which, before its total synthesis, had a history of many partial syntheses that spanned 150 years and included disputes and frustration.

Penicillin V

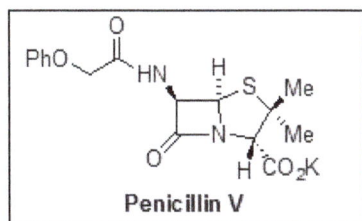

Penicillin V

Penicillins are produced from the mould Penicillium notatum, different strains produce different penicillins. They owe their importance to their powerful effect on various pathogenic organisms. This section will present Sheehan total synthesis of penicillin V .

Retrosynthetic Analysis

The synthetic strategy employed by Sheehan for penicillin synthesis is shown in figure.

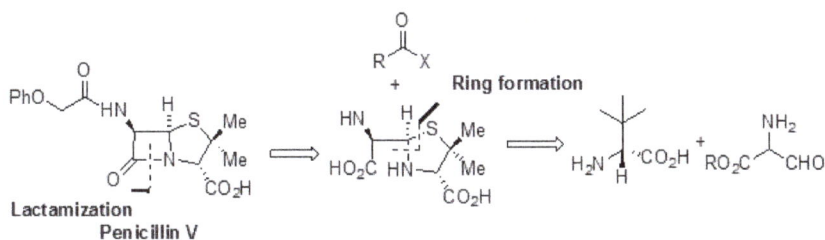

Total Synthesis

Synthesis of (+)-Penicillamine

figure shows the synthesis of (+)-penicillamine from (+)-valine.

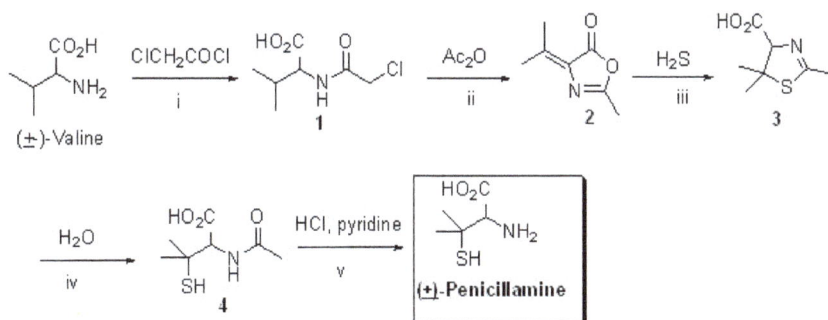

Synthesis of Penicillamine

- N-Acylation of (+)-valine with α -chloroacetyl chloride gives 1 (i) that under-goes dehydrative cyclization with acetic anhydride to provide 2 (ii) by the elim-ination of HCl followed by a proton shift.

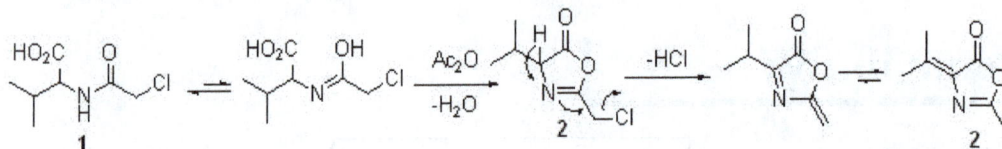

- The azlactone 2 can be cleaved by H$_2$S and the resulting thiol could cyclize via Michael-type addition to the α,β -unsaturated acid to give thiazoline 3 (iii) that could be opened in boiling water to afford 4 (iv). The N-acyl group of 4 could be removed by acid hydrolysis to yield (±)-penicillamine (v) that could be resolved using brucine to give the optically pure (+)-penicillamine.

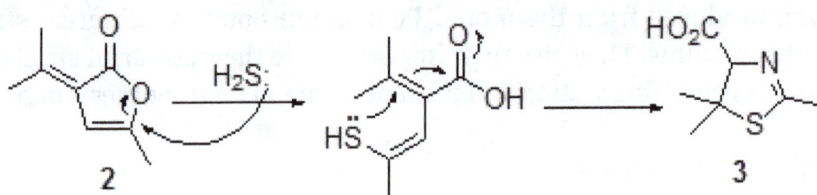

Synthesis of PenicillinV

figure describes the total synthesis of penicillin V.

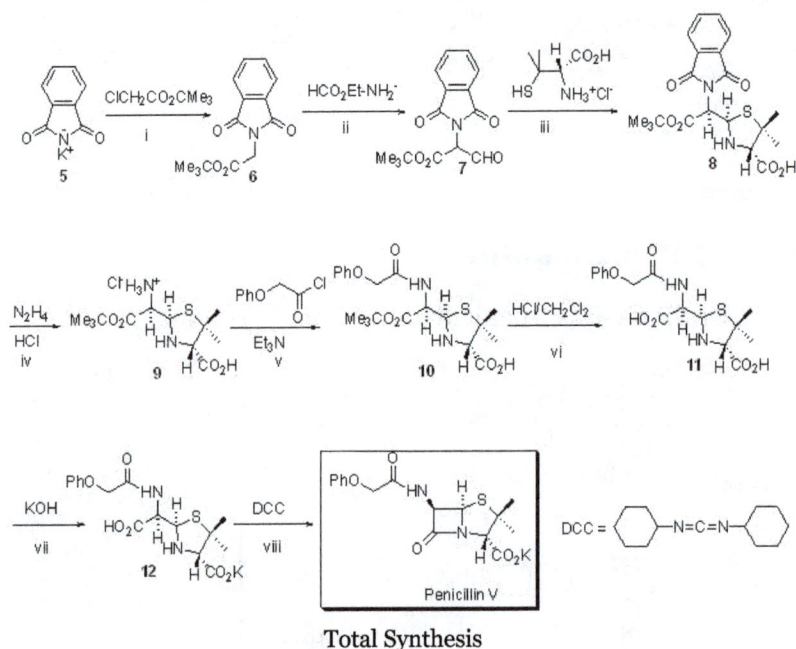

Total Synthesis

- Nucleophilic substitution of 5 with t -butyl chloroacetate gives 6 (Gabriel's synthesis, i) that undergoes cross Claisen condensation with ethyl formate to afford 7 (ii).

- The intermediate 7 with penicillamine hydrochloride at room temperature in sodium acetate buffer affords 8 as a mixture of four diastereomers via Schiff base formation followed by cyclization with the imine double bond. However, the required 8 could be separated from the mixture of diastereomers.

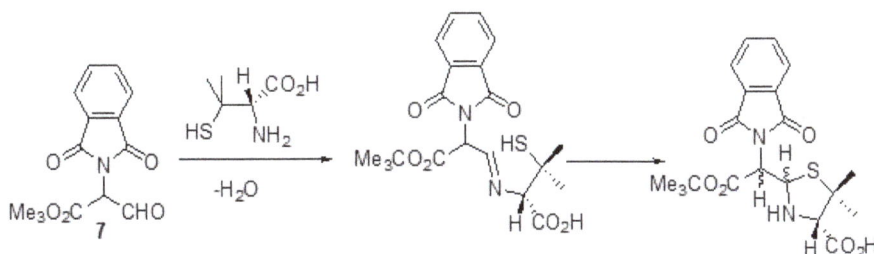

- The removal of the phthalimido group from 8 could be accomplished using hydrazine to afford 9 as a salt in the presence of HCl in acetic acid (iv).

- Acylation of the amino group of 9 in the presence of triethyl amine can give 10 (v) that could be converted into 11 by acid hydrolysis of t-butyl ester in dichloromethane at 0°C (vi).

- The intermediate 11 in the presence of KOH can give the potassium salt (vii) that could be cyclized using DCC to give the potassium penicillinate (viii). Penicillin V can be extracted after acidification with phosphoric acid which crystallizes from aqueous solution at pH 6.8.

Prostaglandin

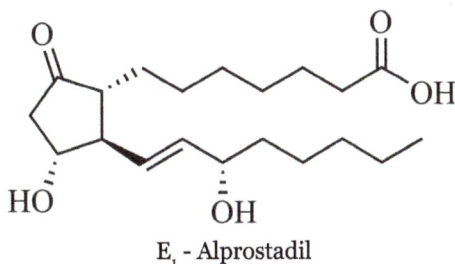

E_1 - Alprostadil

The prostaglandins (PG) are a group of physiologically active lipid compounds having diverse hormone-like effects in animals. Prostaglandins have been found in almost every tissue in humans and other animals. They are derived enzymatically from fatty acids. Every prostaglandin contains 20 carbon atoms, including a 5-carbon ring. They are a subclass of eicosanoids and of the prostanoid class of fatty acid derivatives.

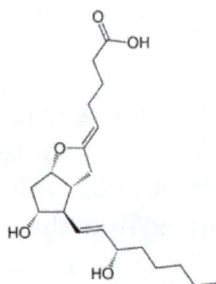

I_2 - Prostacyclin

The structural differences between prostaglandins account for their different biological activities. A given prostaglandin may have different and even opposite effects in different tissues in some cases. The ability of the same prostaglandin to stimulate a reaction in one tissue and inhibit the same reaction in another tissue is determined by the type of receptor to which the prostaglandin binds. They act as autocrine or paracrine factors with their target cells present in the immediate vicinity of the site of their secretion. Prostaglandins differ from endocrine hormones in that they are not produced at a specific site but in many places throughout the human body.

Prostaglandins are powerful locally acting vasodilators and inhibit the aggregation of blood platelets. Through their role in vasodilation, prostaglandins are also involved in inflammation. They are synthesized in the walls of blood vessels and serve the physiological function of preventing needless clot formation, as well as regulating the contraction of smooth muscle tissue. Conversely, thromboxanes (produced by platelet cells) are vasoconstrictors and facilitate platelet aggregation. Their name comes from their role in clot formation (thrombosis).

Specific prostaglandins are named with a letter (which indicates the type of ring structure) followed by a number (which indicates the number of double bonds in the hydrocarbon structure). For example, prostaglandin E1 is abbreviated PGE1 or PGE_1, and prostaglandin I2 is abbreviated PGI2 or PGI_2. The number is traditionally subscripted when the context allow; but, as with many similar subscript-containing nomenclatures, the subscript is simply forgone in many database fields that can store only plain text (such as PubMed bibliographic fields), and readers are used to seeing and writing it without subscript.

History and Name

The name *prostaglandin* derives from the prostate gland. When prostaglandin was first isolated from seminal fluid in 1935 by the Swedish physiologist Ulf von Euler, and independently by M.W. Goldblatt, it was believed to be part of the prostatic secretions. In fact, prostaglandins are produced by the seminal vesicles. It was later shown that many other tissues secrete prostaglandins for various functions. The first total syntheses of prostaglandin $F_{2\alpha}$ and prostaglandin E_2 were reported by E. J. Corey in 1969, an achievement for which he was awarded the Japan Prize in 1989.

In 1971, it was determined that aspirin-like drugs could inhibit the synthesis of prostaglandins. The biochemists Sune K. Bergström, Bengt I. Samuelsson and John R. Vane jointly received the 1982 Nobel Prize in Physiology or Medicine for their research on prostaglandins.

Biochemistry

Biosynthesis

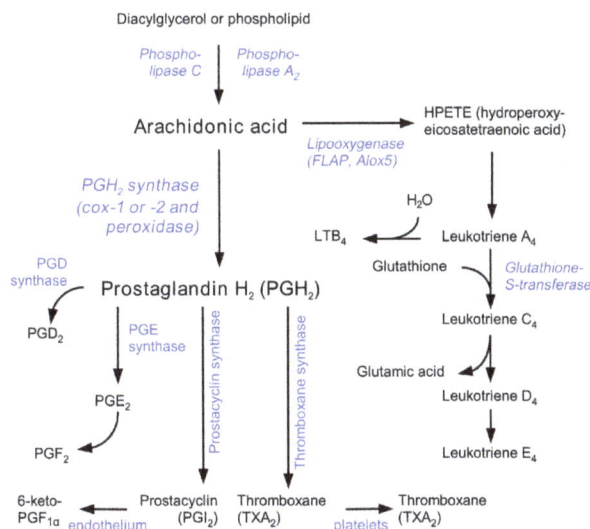

Biosynthesis of eicosanoids

Prostaglandins are found in most tissues and organs. They are produced by almost all nucleated cells. They are autocrine and paracrine lipid mediators that act upon platelets, endothelium, uterine and mast cells. They are synthesized in the cell from the essential fatty acids (EFAs).

An intermediate arachidonic acid is created from diacylglycerol via phospholipase-A$_2$, then brought to either the cyclooxygenase pathway or the lipoxygenase pathway. The cyclooxygenase pathway produces thromboxane, prostacyclin and prostaglandin D, E and F. Alternatively, the lipoxygenase enzyme pathway is active in leukocytes and in macrophages and synthesizes leukotrienes.

Release of Prostaglandins from the Cell

Prostaglandins were originally believed to leave the cells via passive diffusion because of their high lipophilicity. The discovery of the prostaglandin transporter (PGT, SLCO2A1), which mediates the cellular uptake of prostaglandin, demonstrated that diffusion alone cannot explain the penetration of prostaglandin through the cellular membrane. The release of prostaglandin has now also been shown to be mediated by a specific transporter, namely the multidrug resistance protein 4 (MRP4, ABCC4), a member of the ATP-binding cassette transporter superfamily. Whether MRP4 is the only transporter releasing prostaglandins from the cells is still unclear.

Cyclooxygenases

Prostaglandins are produced following the sequential oxidation of arachidonic acid, DGLA or EPA by cyclooxygenases (COX-1 and COX-2) and terminal prostaglandin synthases. The classic dogma is as follows:

- COX-1 is responsible for the baseline levels of prostaglandins.

- COX-2 produces prostaglandins through stimulation.

However, while COX-1 and COX-2 are both located in the blood vessels, stomach and the kidneys, prostaglandin levels are increased by COX-2 in scenarios of inflammation and growth.

Prostaglandin E Synthase

Prostaglandin E_2 (PGE$_2$) is generated from the action of prostaglandin E synthases on prostaglandin H_2 (prostaglandin H2, PGH$_2$). Several prostaglandin E synthases have been identified. To date, microsomal prostaglandin E synthase-1 emerges as a key enzyme in the formation of PGE$_2$.

Other Terminal Prostaglandin Synthases

Terminal prostaglandin synthases have been identified that are responsible for the formation of other prostaglandins. For example, hematopoietic and lipocalin prostaglandin D synthases (hPGDS and lPGDS) are responsible for the formation of PGD$_2$ from PGH$_2$. Similarly, prostacyclin (PGI$_2$) synthase (PGIS) converts PGH$_2$ into PGI$_2$. A thromboxane synthase (TxAS) has also been identified. Prostaglandin-F synthase (PGFS) catalyzes the formation of $9\alpha,11\beta$-PGF$_{2\alpha,\beta}$ from PGD$_2$ and PGF$_{2\alpha}$ from PGH$_2$ in the presence of NADPH. This enzyme has recently been crystallized in complex with PGD$_2$ and bimatoprost (a synthetic analogue of PGF$_{2\alpha}$).

Function

There are currently ten known prostaglandin receptors on various cell types. Prostaglandins ligate a sub-family of cell surface seven-transmembrane receptors, G-protein-coupled receptors. These receptors are termed DP1-2, EP1-4, FP, IP1-2, and TP, corresponding to the receptor that ligates the corresponding prostaglandin (e.g., DP1-2 receptors bind to PGD2).

The diversity of receptors means that prostaglandins act on an array of cells and have a wide variety of effects such as:

- cause constriction or dilation in vascular smooth muscle cells

- cause aggregation or disaggregation of platelets

- sensitize spinal neurons to pain

- induce labor

- decrease intraocular pressure

- regulate inflammation

- regulate calcium movement

- regulate hormones

- control cell growth

- acts on thermoregulatory center of hypothalamus to produce fever

- acts on mesangial cells (specialised smooth muscle cells) in the glomerulus of the kidney to increase glomerular filtration rate

- acts on parietal cells in the stomach wall to inhibit acid secretion

- increase mucus production and bicarbonate secretion

- brain masculinization (in rats)

- increases mating behaviors in goldfish

- Prostaglandins are released during menstruation, due to the destruction of the endometrial cells, and the resultant release of their contents. Release of prostaglandins and other inflammatory mediators in the uterus cause the uterus to contract. These substances are thought to be a major factor in primary dysmenorrhea.

Prostaglandins are potent but have a short half-life before being inactivated and excreted. Therefore, they send only paracrine (locally active) or autocrine (acting on the same cell from which it is synthesized) signals.

Types

The following is a comparison of different types of prostaglandin, prostacyclin I_2 (PGI_2), prostaglandin E_2 (PGE_2), and prostaglandin $F_{2\alpha}$ ($PGF_{2\alpha}$).

Type	Receptor	Receptor type	Function
PGI_2	IP	G_s	• vasodilation • inhibit platelet aggregation • bronchodilation
PGE_2	EP_1	G_q	• bronchoconstriction • GI tract smooth muscle contraction
	EP_2	G_s	• bronchodilation • GI tract smooth muscle relaxation • vasodilation
	EP_3	G_i	• ↓ gastric acid secretion • ↑ gastric mucus secretion • uterus contraction (when pregnant) • GI tract smooth muscle contraction • lipolysis inhibition • ↑ autonomic neurotransmitters • ↑ platelet response to their agonists and ↑ atherothrombosis in vivo
	Unspecified		• hyperalgesia • pyrogenic
$PGF_{2\alpha}$	FP	G_q	• uterus contraction • bronchoconstriction

Role in Pharmacology

Inhibition

Examples of prostaglandin antagonists are:

- NSAIDs (inhibit cyclooxygenase)
- Corticosteroids (inhibit phospholipase A2 production)
- COX-2 selective inhibitors or coxibs
- Cyclopentenone prostaglandins may play a role in inhibiting inflammation

Clinical Uses

Synthetic prostaglandins are used:

- To induce childbirth (parturition) or abortion (PGE_2 or PGF_2, with or without mifepristone, a progesterone antagonist);

- To prevent closure of patent ductus arteriosus in newborns with particular cyanotic heart defects (PGE_1)

- To prevent and treat peptic ulcers (PGE)

- As a vasodilator in severe Raynaud's phenomenon or ischemia of a limb

- In pulmonary hypertension

- In treatment of glaucoma (as in bimatoprost ophthalmic solution, a synthetic prostamide analog with ocular hypotensive activity) ($PGF_{2\alpha}$)

- To treat erectile dysfunction or in penile rehabilitation following surgery (PGE_1 as alprostadil).

- To treat egg binding in small birds

- As an ingredient in eyelash and eyebrow growth beauty products due to side effects associated with increased hair growth

Prostaglandin E$_2$

Prostaglandin E2 (PGE2), also known as dinoprostone, is a naturally occurring prostaglandin which is used as a medication. As a medication it is used in labor induction, bleeding after delivery, termination of pregnancy, and in newborn babies to keep the ductus arteriosus open. In babies it is used in those with congenital heart defects until surgery can be carried out. It may be used within the vagina or by injection into a vein.

Common side effects include vomiting, fever, diarrhea, and excessive uterine contraction. In babies there may be decreased breathing and low blood pressure. Care should be taken in people with asthma or glaucoma and it is not recommended in those who have had a prior C-section. Prostaglandin E2 is in the oxytocics family of medications. It works by binding and activating the prostaglandin E2 receptor which results in the opening and softening of the cervix and dilation of blood vessels.

Prostaglandin E2 was first made in 1970 and approved for medical use in the United States in 1977. It is on the World Health Organization's List of Essential Medicines, the most effective and safe medicines needed in a health system. In the United Kingdom a dose costs the NHS about 8.50 to 30.00 pounds. In the United States a course of treatment costs more than 200 USD. Prostaglandin E2 works as well as prostaglandin E1 in babies; however, is much less expensive.

Physiological Effects

It has important effects in labour (softening the cervix and causing uterine contraction) and also stimulates osteoblasts to release factors that stimulate bone resorption by osteoclasts. PGE2 is also the prostaglandin that ultimately induces fever. It is also implicated in duct-dependent congenital heart diseases and is used in infusion in order to open the duct although PGE1 is more commonly used.

It is a direct vasodilator, relaxing smooth muscles, and it inhibits the release of noradrenaline from sympathetic nerve terminals. It does not inhibit platelet aggregation, where PGI2 does.

PGE2 also suppresses T cell receptor signaling and may play a role in resolution of inflammation. Up-regulation of PGE2 has been implicated as a possible cause of nail clubbing. Furthermore, its postpartal synthesis in newborns is considered as one cause of patent ductus arteriosus.

Side Effects

Common side effects include vomiting, fever, diarrhea, and excessive uterine contraction. In babies there may be decreased breathing and low blood pressure. Care should be taken in people with asthma or glaucoma and it is not recommended in those who have had a prior C-section.

Mechanism of Action

PGE2 is a potent activator of the Wnt signaling pathway. It has been implicated in regulating the developmental specification and regeneration of hematopoietic stem cells through cAMP/PKA activity.

History

It was discovered by Bunting, Gryglewski, Moncada and Vane in 1976.

Society and Culture

It is sold under the trade name of Cervidil(US) and Propess (by Ferring Pharmaceuticals). This is a controlled release vaginal insert. Prostin E_2 (by Pfizer Inc.), and Glandin (by Nabiqasim Pharmaceuticals Pakistan) as a vaginal suppository, to prepare the cervix for labour; it is used to induce labour.

Prostaglandin F2alpha

Prostaglandin $F_{2\alpha}$ (PGF$_{2\alpha}$ in prostanoid nomenclature), pharmaceutically termed dinoprost (INN), is a naturally occurring prostaglandin used in medicine to induce labor and as an abortifacient.

In domestic mammals, it is produced by the uterus when stimulated by oxytocin, in the event that there has been no implantation during the luteal phase. It acts on the corpus luteum to cause luteolysis, forming a corpus albicans and stopping the production of progesterone. Action of $PGF_{2\alpha}$ is dependent on the number of receptors on the corpus luteum membrane.

The $PGF_{2\alpha}$ isoform *8-iso-PGF$_{2\alpha}$* was found in significantly increased amounts in patients with endometriosis, thus being a potential causative link in endometriosis-associated oxidative stress.

Mechanism of Action

$PGF_{2\alpha}$ acts by binding to the prostaglandin F2α receptor.

Synthesis

In 2012 a concise and highly stereoselective total synthesis of $PGF_{2\alpha}$ was described. The synthesis requires only seven steps, a huge improvement on the original 17-steps synthesis of Corey and Cheng, and uses 2,5-dimethoxytetrahydrofuran as a starting reagent, with *S*-proline as an asymmetric catalyst.

Analogues

The following medications are analogues of prostaglandin $F_{2\alpha}$:

- Latanoprost

- Bimatoprost

- Travoprost

- Carboprost

Prostaglandins E_2 and F_{2a}

Prostaglandins are a series of closely related hormones that are derivatives of 'prostanoic acid':

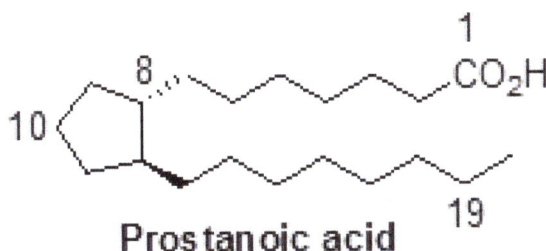

Prostanoic acid

Parent skeleton of the prostaglandin family

Prostaglandins are present in many mammalian tissues at very low concentrations and exhibit potent effects on various types of smooth muscle. They are of considerable medical interest for the control of hypertension.

Prostaglandins E_2 (PGE$_2$) and F_{2a} (PGF$_{2\alpha}$) are two of the six primary prostaglandins. The E series have a β-hydroxy ketone structure in the ring and differ in the degree of unsaturation in the side-chain, while F series have a β-hydroxy group in the ring and likewise differ in the extent of unsaturation in the side-chains.

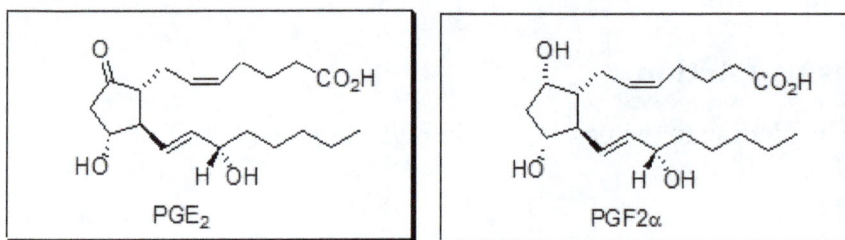

Retrosynthetic Analysis of PGF$_{2a}$

Figure outlines the general features of Corey's strategy for PGF$_{2a}$ synthesis.

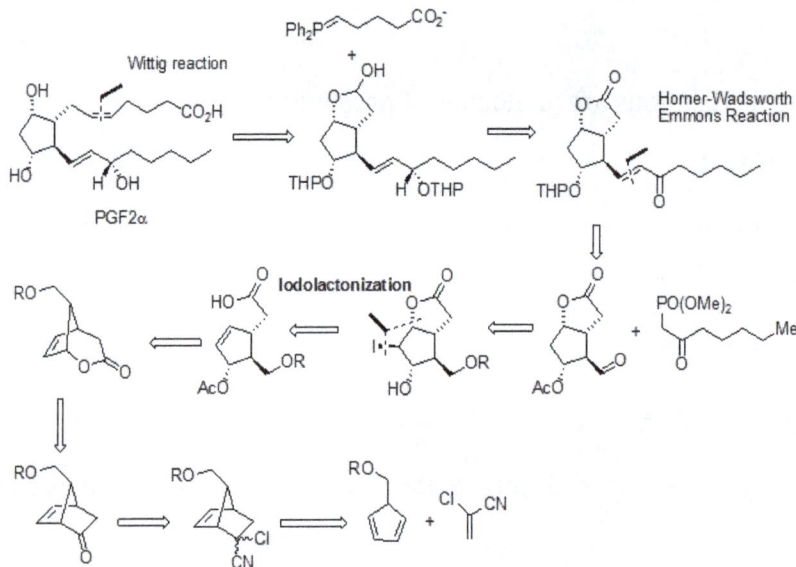

Total Synthesis

figure presents the synthesis of PGE$_2$ and PGF$_{2a}$.

- Thallium(I) cyclopentadiene 1 , prepared from cyclopentadiene with TlSO$_4$ and KOH in water, could be alkylated using benzyl chloromethyl ether, which has the advantage that subsequent debenzylation can be more easily accomplished (i).

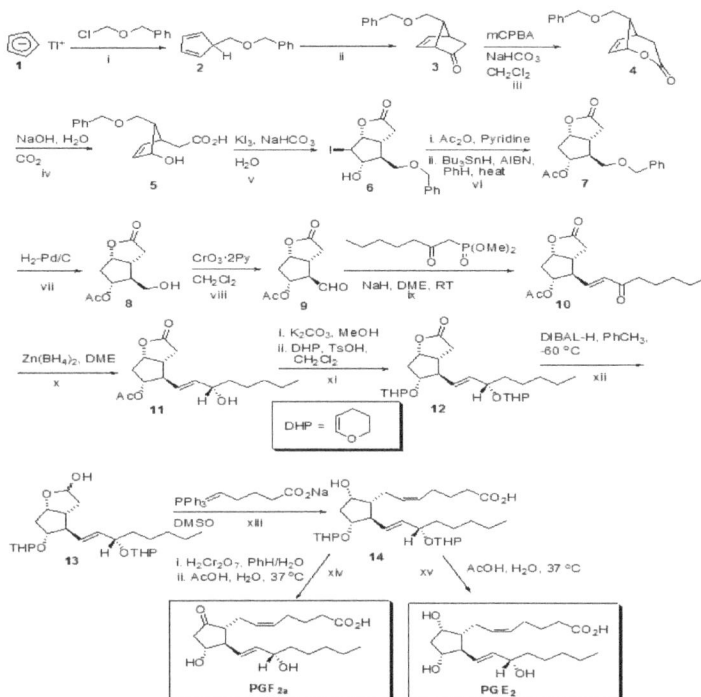

- The copper(II)-catalyzed [4:2] cycloaddition of 2 with 2-chloroacrylonitrile can give 15 that could be hydrolyzed using KOH in DMSO to afford 3 (ii).

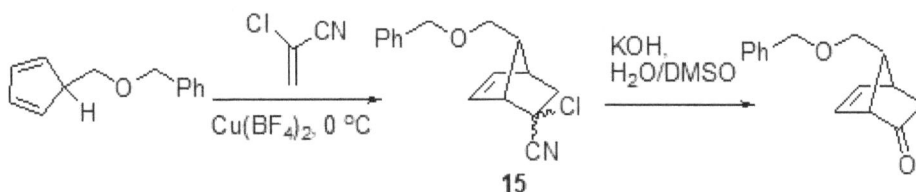

- Baeyer-Villiger oxidation of the ketone 3 can give lactone 4 resulting from migration of secondary carbon in preference to the primary carbon (iii).

- The lactone 4 can be hydrolyzed with aqueous NaOH and the free acid 5 could be obtained by neutralization with CO_2 (iv). The latter could be iodolactonized with KI_3 to afford 6 having five asymmetric centres (v).

- Acetylation of the OH group employing acetic anhydride as an acylating agent followed by deiodination using tributyltin hydride (Bu_3SnH) in the presence of radical initiator AIBN can give 7 (vi) that could be debenzylated by hydrogenolysis to afford 8 (vii).

- The PDC promoted alcohol oxidation of 8 can give the aldehydes 9 (viii) that could undergo Wadsworth-Emmons reaction with the anion of dimethyl 2-oxo-heptyl phosphonate to give 10 (ix).

- The $Zn(BH_4)_2$ mediated reduction of the carbonyl group of the side-chain can yield 11 (x). The protection of OH groups of 11 can be readily accomplished with DHP in the presence of TsOH to afford 12 (xi).

- The selective reduction of 11 to lactol 12 using DIBAL-H (xii) followed by Wittig reaction on the masked aldehydes in DMSO with the phosphorus ylide can afford 13 (xiii).

- The resultant prostanoid material 13 could be converted into prostaglandins E_2 by oxidation of the unprotected hydroxyl group followed by aqueous acidic hydrolysis (xv), whilest the aqueous acid hydrolysis of 13 could afford $F_{2\alpha}$ (xiv).

Ibogamine

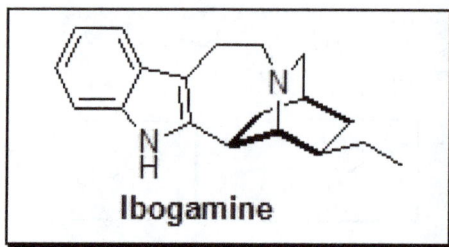

Ibogamine

Ibogamine is an alkaloid in the oboga family that has the clinically important antitumor alkaloid vinblastine. Hence, efficient synthetic approaches for the construction of the family of compounds have been stimulated. This section focuses on a short, stereocontrolled synthesis of ibogamine.

Retrosynthetic Analysis

Figure outlines the common featurs of Trost's strategy for ibogamine synthesis:

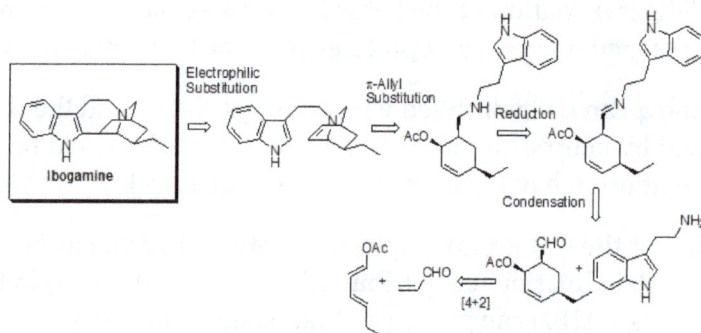

Synthesis of Ibogamine

The synthesis of ibogamine could be accomplished in five steps employing Diels-Alder reaction as key step to control the stereochemistry.

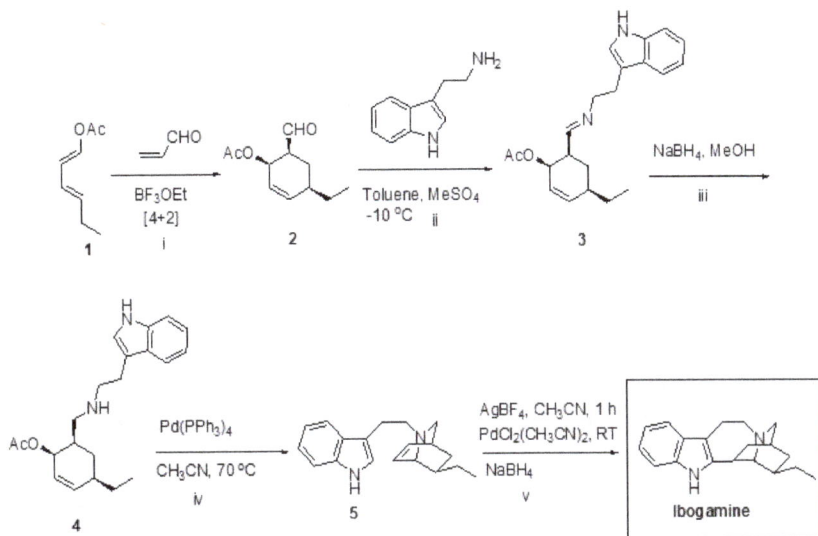

- The Diels-Alder reaction of diene 1 with acrolein could afford the six-membered ring 2 having all the three substituents cis (i). Although, it is immaterial for the acetoxy group, the cis relationship between the ethyl and aldehydes groups is required to obtain the right stereochemistry of the target molecule.

- The Schiff base formation of 2 with tryptamine can give 3 (ii) that could be readily reduced using $NaBH_4$ to afford 4 (iii).

- The intermediate 4 may undergo reaction with Pd(0) to give π -allylic complex with a loss of acetate ion, that could react with the nucleophilic nitrogen atom of the amino group to afford 5 (iv).

- The $PdCl_2$ effected electrophilic substitution of the indole ring can give metal salt complex that could be reduced using $NaBH_4$ to afford the target molecule (v). The presence of silver ion increases the reactivity of the process.

M = silver-palladium salt complex or partially ionized palladium salt

Adenosine Triphosphate

Adenosine triphosphate (ATP) is a nucleotide, also called a nucleoside triphosphate, is a small molecule used in cells as a coenzyme. It is often referred to as the "molecular unit of currency" of intracellular energy transfer.

ATP transports chemical energy within cells for metabolism. Most cellular functions need energy in order to be carried out: synthesis of proteins, synthesis of membranes, movement of the cell, cellular division, transport of various solutes etc. The ATP is the molecule that carries energy to the place where the energy is needed. When ATP breaks into ADP (Adenosine diphosphate) and P_i (phosphate), the breakdown of the last covalent link of phosphate (a simple $-PO_4$) liberates energy that is used in reactions where it is needed.

It is one of the end products of photophosphorylation, aerobic respiration, and fermentation, and is used by enzymes and structural proteins in many cellular processes, including biosynthetic reactions, motility, and cell division. One molecule of ATP contains adenine, ribose, and three phosphate groups, and it is produced by a wide variety of enzymes, including ATP synthase, from adenosine diphosphate (ADP) or adenosine monophosphate (AMP) and various phosphate group donors. Substrate-level phosphorylation, oxidative phosphorylation in cellular respiration, and photophosphorylation in photosynthesis are three major mechanisms of ATP biosynthesis.

Metabolic processes that use ATP as an energy source convert it back into its precursors. ATP is therefore continuously recycled in organisms: the human body, which on average contains only 250 grams (8.8 oz) of ATP, turns over its own body weight equivalent in ATP each day.

ATP is used as a substrate in signal transduction pathways by kinases that phosphorylate proteins and lipids. It is also used by adenylate cyclase, which uses ATP to produce the second messenger molecule cyclic AMP. The ratio between ATP and AMP is used as a way for a cell to sense how much energy is available and control the metabolic pathways that produce and consume ATP. Apart from its roles in signaling and energy metabolism, ATP is also incorporated into nucleic acids by polymerases in the process of transcription. ATP is the neurotransmitter believed to signal the sense of taste.

The structure of this molecule consists of a purine base (adenine) attached by the 9 nitrogen atom to the 1' carbon atom of a pentose sugar (ribose). Three phosphate groups are attached at the 5' carbon atom of the pentose sugar. It is the addition and removal of these phosphate groups that inter-convert ATP, ADP and AMP. When ATP is used in DNA synthesis, the ribose sugar is first converted to deoxyribose by ribonucleotide reductase.

ATP was discovered in 1929 by Karl Lohmann, and independently by Cyrus Fiske and Yellapragada Subbarow of Harvard Medical School, but its correct structure was not determined until some years later. It was proposed to be the intermediary molecule between energy-yielding and energy-requiring reactions in cells by Fritz Albert Lipmann in 1941. It was first artificially synthesized by Alexander Todd in 1948.

Physical and Chemical Properties

ATP consists of adenosine – composed of an adenine ring and a ribose sugar – and three phosphate groups (triphosphate). The phosphoryl groups, starting with the group closest to the ribose, are referred to as the alpha (α), beta (β), and gamma (γ) phosphates. Consequently, it is closely related to the adenosine nucleotide, a monomer of RNA. ATP is highly soluble in water and is quite stable in solutions between pH 6.8 and 7.4, but is rapidly hydrolysed at extreme pH. Consequently, ATP is best stored as an anhydrous salt.

ATP is an unstable molecule in unbuffered water, in which it hydrolyses to ADP and phosphate. This is because the strength of the bonds between the phosphate groups in ATP is less than the strength of the hydrogen bonds (hydration bonds), between its products (ADP and phosphate), and water. Thus, if ATP and ADP are in chemical equilibrium in water, almost all of the ATP will eventually be converted to ADP. A system that is far from equilibrium contains Gibbs free energy, and is capable of doing work. Living cells maintain the ratio of ATP to ADP at a point ten orders of magnitude from equilibrium, with ATP concentrations fivefold higher than the concentration of ADP. This displacement from equilibrium means that the hydrolysis of ATP in the cell releases a large amount of free energy.

Two phosphoanhydride bonds (those that connect adjacent phosphates) in an ATP molecule are responsible for the high energy content of this molecule. In the context of biochemical reactions, these anhydride bonds are frequently – and sometimes controversially – referred to as *high-energy bonds* (despite the fact it takes energy to break bonds). Energy stored in ATP may be released upon hydrolysis of the anhydride bonds. The primary phosphate group on the ATP molecule that is hydrolyzed when energy is needed to drive anabolic reactions is the γ-phosphate group. Located the farthest from the ribose sugar, it has a higher energy of hydrolysis than either the α- or β-phosphate. The bonds formed after hydrolysis – or the phosphorylation of a residue by ATP – are lower in energy than the phosphoanhydride bonds of ATP. During enzyme-catalyzed hydrolysis of ATP or phosphorylation by ATP, the available free energy can be harnessed by a living system to do work.

Any unstable system of potentially reactive molecules could potentially serve as a way of storing free energy, if the cell maintained their concentration far from the equilibrium point of the reaction. However, as is the case with most polymeric biomolecules, the breakdown of RNA, DNA, and ATP into simpler monomers is driven by both energy-release and entropy-increase considerations, in both standard concentrations, and also those concentrations encountered within the cell.

The standard amount of energy released from hydrolysis of ATP can be calculated from the changes in energy under non-natural (standard) conditions, then correcting to biological concentrations. The net change in heat energy (enthalpy) at standard temperature and pressure of the decomposition of ATP into hydrated ADP and hydrated inorganic phosphate is −30.5 kJ/mol, with a change in free energy of 3.4 kJ/mol. The energy released by cleaving either a phosphate (P_i) or pyrophosphate (PP_i) unit from ATP at standard state of 1 M are:

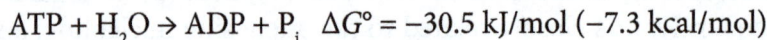

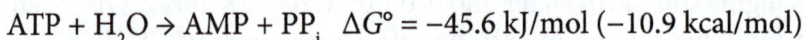

$$ATP + H_2O \rightarrow ADP + P_i \quad \Delta G° = -30.5 \text{ kJ/mol} (-7.3 \text{ kcal/mol})$$

$$ATP + H_2O \rightarrow AMP + PP_i \quad \Delta G° = -45.6 \text{ kJ/mol} (-10.9 \text{ kcal/mol})$$

These values can be used to calculate the change in energy under physiological conditions and the cellular ATP/ADP ratio. However, a more representative value (which takes AMP into consideration) called the Energy charge is increasingly being employed. The values given for the Gibbs free energy for this reaction are dependent on a number of factors, including overall ionic strength and the presence of alkaline earth metal ions such as Mg^{2+} and Ca^{2+}. Under typical cellular conditions, ΔG is approximately −57 kJ/mol (−14 kcal/mol).

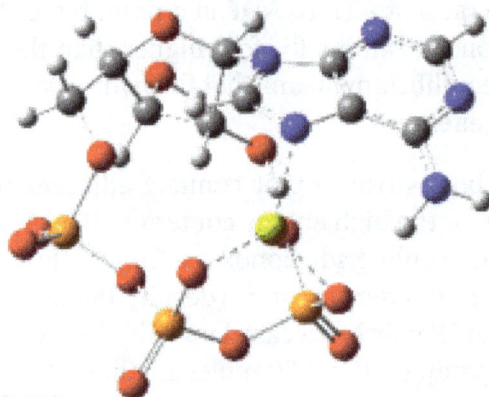

This image shows a full 360-degree rotation of a single, gas-phase magnesium-ATP chelate with a charge of −2. The molecule was optimized at the UB3LYP/6-311++G(d,p) theoretical level and the atomic connectivity modified by the human optimizer to reflect the probable electronic structure.

Ionization in Biological Systems

ATP (adenosine triphosphate) has multiple groups with different acid dissociation constants. In neutral solution, ionized ATP exists mostly as ATP^{4-}, with a small proportion of ATP^{3-}. As ATP has several negatively charged groups in neutral solution, it can chelate metals with very high affinity. The binding constant for various metal ions are (given as per mole) as Mg^{2+} (9554), Na^+ (13), Ca^{2+} (3722), K^+ (8), Sr^{2+} (1381) and Li^+ (25). Due to the strength of these interactions, ATP exists in the cell mostly in a complex with Mg^{2+}.

Biosynthesis

The ATP concentration inside the cell is typically 1–10 mM. ATP can be produced by redox reactions using simple and complex sugars (carbohydrates) or lipids as an energy source. For complex fuels to be synthesized into ATP, they first need to be broken down into smaller, more simple molecules. Carbohydrates are hydrolysed into simple sugars, such as glucose and fructose. Fats (triglycerides) are metabolised to give fatty acids and glycerol.

The overall process of oxidizing glucose to carbon dioxide is known as cellular respiration and can produce about 30 molecules of ATP from a single molecule of glucose. ATP can be produced by a number of distinct cellular processes; the three main pathways used to generate energy in eukaryotic organisms are glycolysis and the citric acid cycle/oxidative phosphorylation, both components of cellular respiration; and beta-oxidation. The majority of this ATP production by a non-photosynthetic aerobic eukaryote takes place in the mitochondria, which can make up nearly 25% of the total volume of a typical cell.

Glycolysis

In glycolysis, glucose and glycerol are metabolized to pyruvate via the glycolytic pathway. In most organisms, this process occurs in the cytosol, but, in some protozoa such as the kinetoplastids, this is carried out in a specialized organelle called the glycosome. Glycolysis generates a net two molecules of ATP through substrate phosphorylation catalyzed by two enzymes: PGK and pyruvate kinase. Two molecules of NADH are also produced, which can be oxidized via the electron transport chain and result in the generation of additional ATP by ATP synthase. The pyruvate generated as an end-product of glycolysis is a substrate for the Krebs Cycle.

Glucose

In the mitochondrion, pyruvate is oxidized by the pyruvate dehydrogenase complex to the acetyl group, which is fully oxidized to carbon dioxide by the citric acid cycle (also known as the Krebs cycle). Every "turn" of the citric acid cycle produces two molecules of carbon dioxide, one molecule of the ATP equivalent guanosine triphosphate (GTP) through substrate-level phosphorylation catalyzed by succinyl-CoA synthetase, three molecules of the reduced coenzyme NADH, and one molecule of the reduced coenzyme $FADH_2$. Both of these latter molecules are recycled to their oxidized states (NAD^+ and FAD, respectively) via the electron transport chain, which generates additional ATP by oxidative phosphorylation. The oxidation of an NADH molecule results in the synthesis of 2–3 ATP molecules, and the oxidation of one $FADH_2$ yields between 1–2 ATP molecules. The majority of cellular ATP is generated by this process. Although the citric acid cycle itself does not involve molecular oxygen, it is an obligately aerobic process because O_2 is needed to recycle the reduced NADH and $FADH_2$ to their oxidized states. In the absence of oxygen the citric acid cycle will cease to function due to the lack of available NAD^+ and FAD.

The generation of ATP by the mitochondrion from cytosolic NADH relies on the malate-aspartate shuttle (and to a lesser extent, the glycerol-phosphate shuttle) because the inner mitochondrial membrane is impermeable to NADH and NAD$^+$. Instead of transferring the generated NADH, a malate dehydrogenase enzyme converts oxaloacetate to malate, which is translocated to the mitochondrial matrix. Another malate dehydrogenase-catalyzed reaction occurs in the opposite direction, producing oxaloacetate and NADH from the newly transported malate and the mitochondrion's interior store of NAD$^+$. A transaminase converts the oxaloacetate to aspartate for transport back across the membrane and into the intermembrane space.

In oxidative phosphorylation, the passage of electrons from NADH and FADH$_2$ through the electron transport chain powers the pumping of protons out of the mitochondrial matrix and into the intermembrane space. This creates a proton motive force that is the net effect of a pH gradient and an electric potential gradient across the inner mitochondrial membrane. Flow of protons down this potential gradient – that is, from the intermembrane space to the matrix – provides the driving force for ATP synthesis by ATP synthase. This enzyme contains a rotor subunit that physically rotates relative to the static portions of the protein during ATP synthesis.

Most of the ATP synthesized in the mitochondria will be used for cellular processes in the cytosol; thus it must be exported from its site of synthesis in the mitochondrial matrix. The inner membrane contains an antiporter, the ADP/ATP translocase, which is an integral membrane protein used to exchange newly synthesized ATP in the matrix for ADP in the intermembrane space. This translocase is driven by the membrane potential, as it results in the movement of about 4 negative charges out of the mitochondrial membrane in exchange for 3 negative charges moved inside. However, it is also necessary to transport phosphate into the mitochondrion; the phosphate carrier moves a proton in with each phosphate, partially dissipating the proton gradient.

Beta Oxidation

Fatty acids can also be broken down to acetyl-CoA by beta-oxidation. Each round of this cycle reduces the length of the acyl chain by two carbon atoms and produces one NADH and one FADH$_2$ molecule, which are used to generate ATP by oxidative phosphorylation. Because NADH and FADH$_2$ are energy-rich molecules, dozens of ATP molecules can be generated by the beta-oxidation of a single long acyl chain. The high energy yield of this process and the compact storage of fat explain why it is the most dense source of dietary calories.

Fermentation

Fermentation entails the generation of energy via the process of substrate-level phosphorylation in the absence of a respiratory electron transport chain. In most eukaryotes, glucose is used as both an energy store and an electron donor. The equation for the oxidation of glucose to lactic acid is:

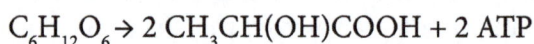

$$C_6H_{12}O_6 \rightarrow 2\ CH_3CH(OH)COOH + 2\ ATP$$

Anaerobic Respiration

Anaerobic respiration is the process of respiration using an electron acceptor other than O_2. In prokaryotes, multiple electron acceptors can be used in anaerobic respiration. These include nitrate, sulfate or carbon dioxide. These processes lead to the ecologically important processes of denitrification, sulfate reduction and acetogenesis, respectively.

ATP Replenishment by Nucleoside Diphosphate Kinases

ATP can also be synthesized through several so-called "replenishment" reactions catalyzed by the enzyme families of nucleoside diphosphate kinases (NDKs), which use other nucleoside triphosphates as a high-energy phosphate donor, and the ATP:guanido-phosphotransferase family.

ATP Production During Photosynthesis

In plants, ATP is synthesized in thylakoid membrane of the chloroplast during the light-dependent reactions of photosynthesis in a process called photophosphorylation. Here, light energy is used to pump protons across the chloroplast membrane. This produces a proton-motive force and this drives the ATP synthase, exactly as in oxidative phosphorylation. Some of the ATP produced in the chloroplasts is consumed in the Calvin cycle, which produces triose sugars.

ATP Recycling

The total quantity of ATP in the human body is about 0.2 moles. The majority of ATP is not usually synthesised *de novo*, but is generated from ADP by the aforementioned processes. Thus, at any given time, the total amount of ATP + ADP remains fairly constant.

The energy used by human cells requires the hydrolysis of 100 to 150 moles of ATP daily, which is around 50 to 75 kg. A human will typically use up his or her body weight of ATP over the course of the day. This means that each ATP molecule is recycled 500 to 750 times during a single day (100 / 0.2 = 500). ATP cannot be stored, hence its consumption closely follows its synthesis. However a total of around 5 g of ATP is used by cell processes at any time in the body.

Regulation of Biosynthesis

ATP production in an aerobic eukaryotic cell is tightly regulated by allosteric mechanisms, by feedback effects, and by the substrate concentration dependence of individual enzymes within the glycolysis and oxidative phosphorylation pathways. Key control points occur in enzymatic reactions that are so energetically favorable that they are effectively irreversible under physiological conditions.

In glycolysis, hexokinase is directly inhibited by its product, glucose-6-phosphate, and pyruvate kinase is inhibited by ATP itself. The main control point for the glycolytic pathway is phosphofructokinase (PFK), which is allosterically inhibited by high concentrations of ATP and activated by high concentrations of AMP. The inhibition of PFK by ATP is unusual, since ATP is also a substrate in the reaction catalyzed by PFK; the biologically active form of the enzyme is a tetramer that exists in two possible conformations, only one of which binds the second substrate fructose-6-phosphate (F6P). The protein has two binding sites for ATP – the active site is accessible in either protein conformation, but ATP binding to the inhibitor site stabilizes the conformation that binds F6P poorly. A number of other small molecules can compensate for the ATP-induced shift in equilibrium conformation and reactivate PFK, including cyclic AMP, ammonium ions, inorganic phosphate, and fructose-1,6- and -2,6-biphosphate.

The citric acid cycle is regulated mainly by the availability of key substrates, particularly the ratio of NAD+ to NADH and the concentrations of calcium, inorganic phosphate, ATP, ADP, and AMP. Citrate – the molecule that gives its name to the cycle – is a feedback inhibitor of citrate synthase and also inhibits PFK, providing a direct link between the regulation of the citric acid cycle and glycolysis.

In oxidative phosphorylation, the key control point is the reaction catalyzed by cytochrome c oxidase, which is regulated by the availability of its substrate – the reduced form of cytochrome c. The amount of reduced cytochrome c available is directly related to the amounts of other substrates:

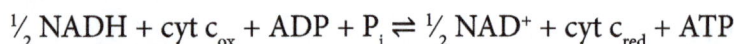

$$\tfrac{1}{2}\,NADH + cyt\,c_{ox} + ADP + P_i \rightleftharpoons \tfrac{1}{2}\,NAD^+ + cyt\,c_{red} + ATP$$

which directly implies this equation:

$$\frac{[cyt\,c_{red}]}{[cyt\,c_{ox}]} = \left(\frac{[NADH]}{[NAD]^+}\right)^{\frac{1}{2}} \left(\frac{[ADP][P_i]}{[ATP]}\right) K_{eq}$$

Thus, a high ratio of [NADH] to [NAD+] or a high ratio of [ADP][P_i] to [ATP] imply a high amount of reduced cytochrome c and a high level of cytochrome c oxidase activity. An additional level of regulation is introduced by the transport rates of ATP and NADH between the mitochondrial matrix and the cytoplasm.

Functions in Cells

Metabolism, Synthesis, and Active Transport

ATP is consumed in the cell by energy-requiring (endergonic) processes and can be generated by energy-releasing (exergonic) processes. In this way ATP transfers energy between spatially separate metabolic reactions. ATP is the main energy source for the majority of cellular functions. This includes the synthesis of macromolecules, including

DNA and RNA, and proteins. ATP also plays a critical role in the transport of macro-molecules across cell membranes, e.g. exocytosis and endocytosis.

Roles in Cell Structure and Locomotion

ATP is critically involved in maintaining cell structure by facilitating assembly and dis-assembly of elements of the cytoskeleton. In a related process, ATP is required for the shortening of actin and myosin filament crossbridges required for muscle contraction. This latter process is one of the main energy requirements of animals and is essential for locomotion and respiration.

Cell Signalling

Extracellular Signalling

Extracellular ATP (eATP) is also a signalling molecule. ATP, ADP, or adenosine are recognised by purinergic receptors, or purinoreceptors, which might be the most abundant receptors in mammalian tissues.

In humans, this signalling role is important in both the central and peripheral nervous system. Activity-dependent release of ATP from synapses, axons and glia activates purinergic membrane receptors known as P2. The *P2Y* receptors are G protein-coupled receptors, which are *metabotropic*, and primarily modulate intracellular calcium and, to a lesser extent, cyclic AMP levels. Though named between $P2Y_1$ and $P2Y_{15}$, only nine members of the P2Y family have been cloned, and some are only related through weak homology and several ($P2Y_5$, $P2Y_7$, $P2Y_9$, $P2Y_{10}$) do not function as receptors that raise cytosolic calcium. The *P2X ionotropic* receptor subgroup comprises seven members ($P2X_1$–$P2X_7$), which are ligand-gated Ca^{2+} -permeable ion channels that open when bound to an extracellular purine nucleotide, like ATP. In contrast to P2 receptors (agonist rank order of potency: ATP > ADP > AMP > ADO), purinergic nucleoside triphosphates like ATP are not strong agonists of P1 receptors, which are strongly activated by adenosine and other nucleosides (ADO > AMP > ADP > ATP). P1 receptors have A1, A2a, A2b, and A3 subtypes (the "A" is standard nomenclature for indicating an *adenosine receptor* subtype), all of which are G protein-coupled receptors, A1 and A3 being coupled to Gi, and A2a and A2b being coupled to Gs. All adenosine receptors were shown to activate at least one subfamily of mitogen-activated protein kinases. The actions of adenosine are often antagonistic or synergistic to the actions of ATP. In the CNS, adenosine has multiple functions, such as modulation of neural development, neuron and glial signalling and the control of innate and adaptive immune systems.

Intracellular Signaling

ATP is critical in signal transduction processes. It is used by kinases as the source of phosphate groups in their phosphate transfer reactions. Kinase activity on substrates such as proteins or membrane lipids are a common form of signal transduction. Phosphorylation of a protein by a kinase can activate this cascade such as the mitogen-activated protein kinase cascade.

ATP is also used by adenylate cyclase most commonly in G protein-coupled receptor signal transduction pathways and is transformed to the second messenger molecule cyclic AMP, which is involved in triggering calcium signals by the release of calcium from intracellular stores. This form of signal transduction is particularly important in brain function, although it is involved in the regulation of a multitude of other cellular processes.

DNA and RNA Synthesis

In all known organisms, the Deoxyribonucleotides that make up DNA are synthesized by the action of ribonucleotide reductase (RNR) enzymes on their corresponding ribonucleotides. These enzymes reduce the sugar residue from ribose to deoxyribose by removing oxygen from the 2′ hydroxyl group; the substrates are ribonucleoside diphosphates and the products deoxyribonucleoside diphosphates (the latter are denoted dADP, dCDP, dGDP, and dUDP respectively.) All ribonucleotide reductase enzymes use a common sulfhydryl radical mechanism reliant on reactive cysteine residues that oxidize to form disulfide bonds in the course of the reaction. RNR enzymes are recycled by reaction with thioredoxin or glutaredoxin.

The regulation of RNR and related enzymes maintains a balance of dNTPs relative to each other and relative to NTPs in the cell. Very low dNTP concentration inhibits DNA synthesis and DNA repair and is lethal to the cell, while an abnormal ratio of dNTPs is mutagenic due to the increased likelihood of the DNA polymerase incorporating the wrong dNTP during DNA synthesis. Regulation of or differential specificity of RNR has been proposed as a mechanism for alterations in the relative sizes of intracellular dNTP pools under cellular stress such as hypoxia.

In the synthesis of the nucleic acid RNA, adenosine derived from ATP is one of the four nucleotides incorporated directly into RNA molecules by RNA polymerases. The energy driving this polymerization comes from cleaving off a pyrophosphate (two phosphate groups). The process is similar in DNA biosynthesis, except that ATP is reduced to the deoxyribonucleotide dATP, before incorporation into DNA.

Amino Acid Activation in Protein Synthesis

Aminoacyl-tRNA synthetase enzymes utilize ATP as an energy source to attach a tRNA molecule to its specific amino acid, forming an aminoacyl-tRNA complex, ready for translation at ribosomes. The energy is made available by ATP hydrolysis to adenosine monophosphate (AMP) as two phosphate groups are removed. Amino acid activation refers to the attachment of an amino acid to its Transfer RNA (tRNA). Aminoacyl transferase binds Adenosine triphosphate (ATP) to amino acid, PP is released. Aminoacyl transferase binds AMP-amino acid to tRNA. The AMP is used in this step.

Amino Acid Activation

During amino acid activation the amino acids (aa) are attached to their corresponding tRNA. The coupling reactions are catalysed by a group of enzymes called aminoacyl-tRNA synthetases (named after the reaction product aminoacyl-tRNA or aa-tRNA). The coupling reaction proceeds in two steps:

1. aa + ATP --> aa-AMP + PP$_i$

2. aa-AMP + tRNA --> aa-tRNA + AMP

The amino acid is coupled to the penultimate nucleotide at the 3'-end of the tRNA (the A in the sequence CCA) via an ester bond. The formation of the ester bond conserves a considerable part of the energy from the activation reaction. This stored energy provides the majority of the energy needed for peptide bond formation during translation.

Each of the 20 amino acids are recognized by its specific aminoacyl-tRNA synthetase. The synthetases are usually composed of one to four protein subunits. The enzymes vary considerably in structure although they all perform the same type of reaction by binding ATP, one specific amino acid and its corresponding tRNA.

The specificity of the amino acid activation is as critical for the translational accuracy as the correct matching of the codon with the anticodon. The reason is that the ribosome only sees the anticodon of the tRNA during translation. Thus, the ribosome will not be able to discriminate between tRNAs with the same anticodon but linked to different amino acids.

The error frequency of the amino acid activation reaction is approximately 1 in 10000 despite the small structural differences between some of the amino acids.

Binding to Proteins

Some proteins that bind ATP do so in a characteristic protein fold known as the Rossmann fold, which is a general nucleotide-binding structural domain that can also bind the coenzyme NAD. The most common ATP-binding proteins, known as kinases, share a small number of common folds; the protein kinases, the largest kinase superfamily, all share common structural features specialized for ATP binding and phosphate transfer.

ATP in complexes with proteins, in general, requires the presence of a divalent cation, almost always magnesium, which binds to the ATP phosphate groups. The presence of magnesium greatly decreases the dissociation constant of ATP from its protein binding partner without affecting the ability of the enzyme to catalyze its reaction once the ATP has bound. The presence of magnesium ions can serve as a mechanism for kinase regulation.

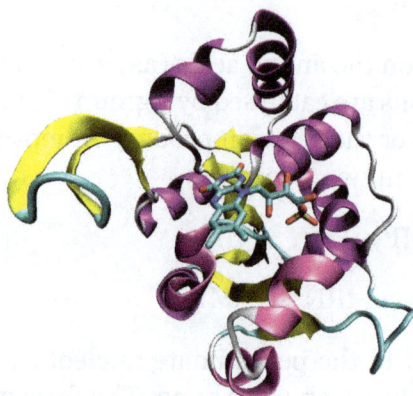

An example of the Rossmann fold, a structural domain of a decarboxylase enzyme from the bacterium *Staphylococcus epidermidis* (PDB: 1G5Q) with a bound flavin mononucleotide cofactor.

ATP Analogues

Biochemistry laboratories often use *in vitro* studies to explore ATP-dependent molecular processes. Enzyme inhibitors of ATP-dependent enzymes such as kinases are needed to examine the binding sites and transition states involved in ATP-dependent reactions. ATP analogs are also used in X-ray crystallography to determine a protein structure in complex with ATP, often together with other substrates. Most useful ATP analogs cannot be hydrolyzed as ATP would be; instead they trap the enzyme in a structure closely related to the ATP-bound state. Adenosine 5'-(γ-thiotriphosphate) is an extremely common ATP analog in which one of the gamma-phosphate oxygens is replaced by a sulfur atom; this molecule is hydrolyzed at a dramatically slower rate than ATP itself and functions as an inhibitor of ATP-dependent processes. In crystallographic studies, hydrolysis transition states are modeled by the bound vanadate ion. However, caution is warranted in interpreting the results of experiments using ATP analogs, since some enzymes can hydrolyze them at appreciable rates at high concentration.

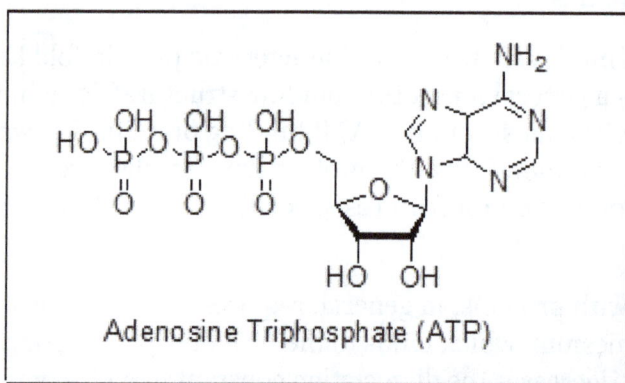

Adenosine Triphosphate (ATP)

ATP, in conjunction with its diphosphate, ADP, acts as a reversible phosphorylating couple and energy store. For example, it is concerned with the supply of energy for muscular contraction.

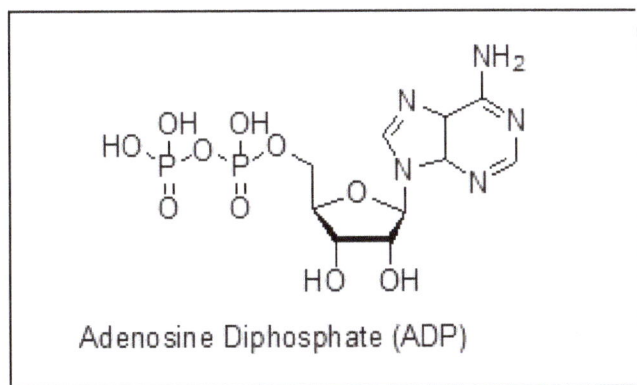

Adenosine Diphosphate (ADP)

Synthesis of β -Chloro-2,3,5-Triacetyl-D-Ribofuranose

The synthesis of β-chloro-2,3,5-triacetyl-D-ribofurance from D-ribose could be accomplished in five steps.

- The selective reaction of the primary OH group of D-ribose with triphenylmethyl chloride ensures the sugar to adopt the furanose ring 1 system (i) that could be transformed into 3 via 2 by acetylation (ii) followed by removel of the triphenylmethyl group of using hydrogenation (iii).

- Acetylation (iv) followed by S_N2 reaction at the carbon next to the ring oxygen of 4 using HCl (v) give the target molecule.

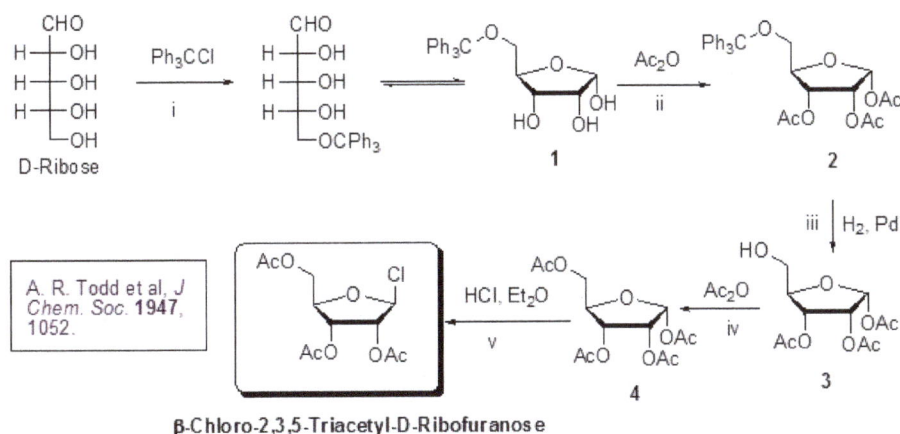

β-Chloro-2,3,5-Triacetyl-D-Ribofuranose

Synthesis of A Denosine

Uric acid could also be converted into adenosine in five steps.

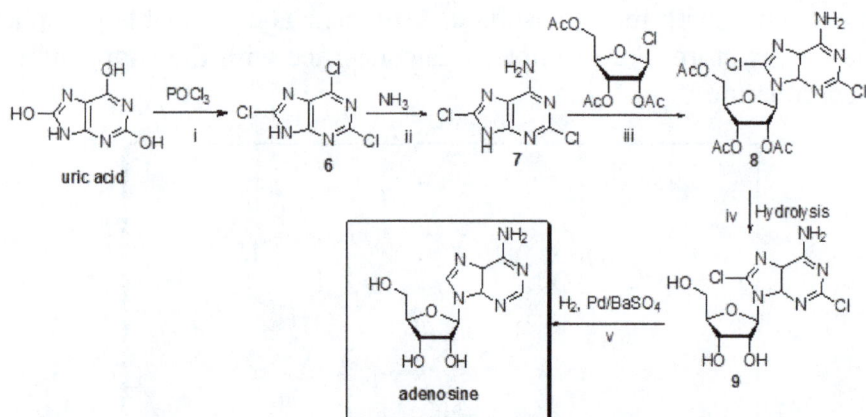

uric acid

adenosine

- Uric acid could be converted into 6 by reaction with $POCl_3$ (i) that could undergo nuleophilic substitution selectively at 6-position with NH_3 to afford 7 (ii).

- Reaction of 7 with the chloro-furanoside gives 8 , presumably as a result of the formation of an acetoxonium ion followed by an S_N2 reaction (iii).

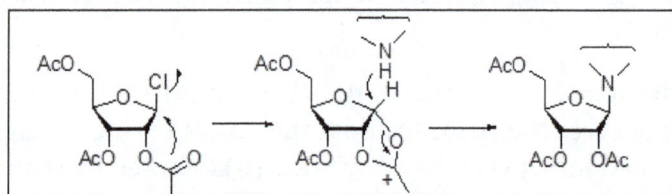

- Hydrolysis of the acetyl groups of 8 (iv) followed by hydrogenolysis of the C-Cl bonds of 9 (v) gives the target adenosine.

Synthesis of ATP

The synthesis of ATP can be accomplished in eight steps form the above synthesized adenosine.

- The 2- and 3-hydroxyl groups of the furanose ring of adenosine could be protected by the formation of ketal to afford ketal 10 (i).

- Phosphorylation of 10 could be effected to give 11 at a low temperature and the removal of HCl can be performed using pyridine as a solvent (ii).

- Mild acid hydrolysis of 11 leads to the formation of 12 by removal of the one of benzyl groups along with the isopropylidene group (iii). After removal of the acid as barium sulfate, the product could be dissolved in alkali and precipitated as its silver salt.

- Phosphorylation of 12 in anhydrous CH_3COOH can give 13 (iv) that could be selectively debenzylated with N -methylmorpholine to afford 14 (v).

- Treatment of 14 with AgNO$_3$ can give 15 (vi) that could be phosphorylated in a mixture of CH$_3$CN and PhOH to afford 16 (vii).

- The four benzyl groups of 16 could be removed by hydrogenolysis and the target product, ATP, can be precipitated as its barium salt, liberated with sulfuric acid, and isolated as its acridinium salt (viii).

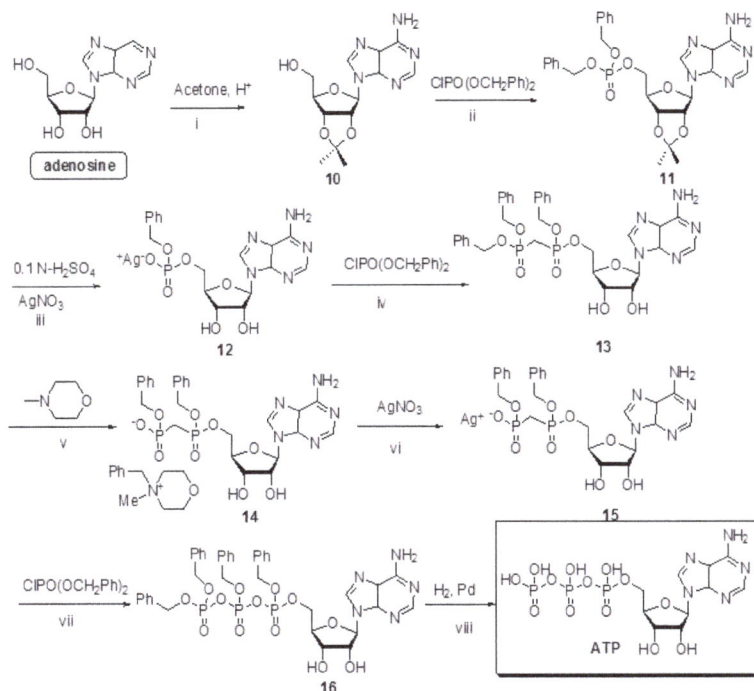

Preparation of Dibenzyl Chlorophosphonate

The chlorination of dibenzyl phosphate can give the phosphorylating agent, dibenzyl chlorophosphonate, in carbon tetrachloride.

$$2PhCH_2OH + H_3PO_3 \longrightarrow HPO(OCH_2Ph)_2 \xrightarrow{Cl_2} ClPO(OCH_2Ph)_2$$

References

- Baumeister, A. A.; Hawkins, M. F.; Uzelac, S. M. (2003). "The Myth of Reserpine-Induced Depression: Role in the Historical Development of the Monoamine Hypothesis". Journal of the History of the Neurosciences. 12 (2): 207–220. PMID 12953623. doi:10.1076/jhin.12.2.207.15535

- Ramawat et al, 1999.Ramawat KG; Rachnana Sharma; Suri SS. Ramawat, KG.; Merillon, JM., eds. Medicinal Plants in Biotechnology- Secondary metabolites 2nd edition 2007. Oxford and IBH, India. pp. 66–367. ISBN 978-1-57808-428-9

- WHO Model Formulary 2008 (PDF). World Health Organization. 2009. p. X. ISBN 9789241547659. Retrieved 8 December 2016

- E. J. Corey (1988). "Retrosynthetic Thinking - Essentials and Examples". Chem. Soc. Rev. 17: 111–133. doi:10.1039/CS9881700111

- Lemieux G, Davignon A, Genest J (1956). "Depressive states during Rauwolfia therapy for arterial hypertension; a report of 30 cases". Canadian Medical Association Journal. 74 (7): 522–6. PMC 1823144. PMID 13304797

- Nicolaou, K. C.; E. J. Sorensen (1996). Classics in Total Synthesis. Weinheim, Germany: VCH. p. 55. ISBN 3-527-29284-5

- "WHO Model List of Essential Medicines (19th List)" (PDF). World Health Organization. April 2015. Retrieved 8 December 2016

- Von Euler US (1935). "Über die spezifische blutdrucksenkende Substanz des menschlichen Prostata- und Samenblasensekrets" (PDF). Wien Klin Wochenschr. 14 (33): 1182–3. doi:10.1007/BF01778029

- James Law et.al:"Route Designer: A Retrosynthetic Analysis Tool Utilizing Automated Retrosynthetic Rule Generation", Journal of Chemical Information and Modelling (ACS JCIM) Publication Date (Web): February 6, 2009; doi:10.1021/ci800228y

- AJ Giannini,HR Black. Psychiatric, Psychogenic, and Somatopsychic Disorders Handbook. Garden City,NY. Medical Examination Publishing, 1978. Pg. 233. ISBN 0-87488-596-5

- "Penicillin V Potassium tablet: Drug Label Sections". U.S. National Library of Medicine, Daily Med: Current Medication Information. December 2006. Retrieved 2009-08-02

- Coulthard, G.; Erb, W.; Aggarwal, V. K. (2012). "Stereocontrolled organocatalytic synthesis of prostaglandin PGF2α in seven steps". Nature. 489 (7415): 278–281. PMID 22895192. doi:10.1038/nature11411

- Du Vigneaud V, Ressler C, Swan JM, Roberts CW, Katsoyannis PG (1954). "The Synthesis of Oxytocin". Journal of the American Chemical Society. 76 (12): 3115–3121. doi:10.1021/ja01641a004

- Hamilton, Richart (2015). Tarascon Pocket Pharmacopoeia 2015 Deluxe Lab-Coat Edition. Jones & Bartlett Learning. p. 95. ISBN 9781284057560

Base Catalyzed Reactions: An Overview

The chapter strategically encompasses and incorporates the major components and key concepts of base catalyzed carbon-carbon bond formation, providing a complete understanding. The formation of a base catalyzed carbon-carbon bond depends on the bond-bond carbon formation derived from organometallic reagents. This chapter discusses the subject matter in a critical manner.

Base Catalyzed Reactions

Principles

The base catalyzed carbon-carbon bond formation is closely related to the carbon-carbon bond formation from organometallic reagents. In both methods, the negatively polarized carbon reacts with electrophilic carbon of carbonyl groups and related compounds.

Reactions of Enolates with Carbonyl Compounds

Aldol Condensation

An aldol condensation is a condensation reaction in organic chemistry in which an enol or an enolate ion reacts with a carbonyl compound to form a β-hydroxyaldehyde or β-hydroxyketone (an aldol reaction), followed by dehydration to give a conjugated enone.

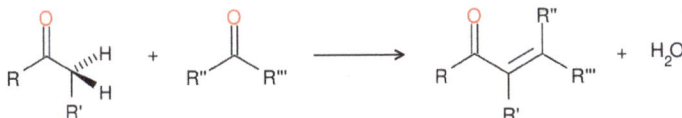

Aldol condensations are important in organic synthesis, because they provide a good way to form carbon–carbon bonds. For example, the Robinson annulation reaction sequence features an aldol condensation; the Wieland-Miescher ketone product is an important starting material for many organic syntheses. Aldol condensations are also commonly discussed in university level organic chemistry classes as a good bond-forming reaction that demonstrates important reaction mechanisms. In its usual form, it involves the nucleophilic addition of a ketone enolate to an aldehyde to form a β-hydroxy ketone, or "aldol" (aldehyde + alcohol), a structural unit found in many naturally occurring molecules and pharmaceuticals.

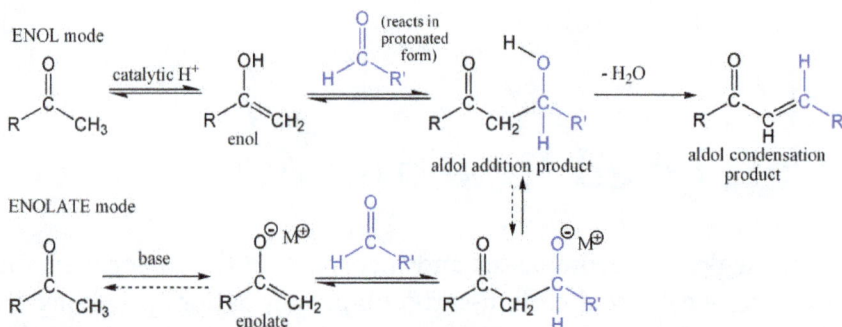

The name aldol condensation is also commonly used, especially in biochemistry, to refer to just the first (addition) stage of the process—the aldol reaction itself—as catalyzed by aldolases. However, the aldol reaction is not formally a condensation reaction because it does not involve the loss of a small molecule.

The reaction between an aldehyde/ketone and an aromatic carbonyl compound lacking an alpha-hydrogen (cross aldol condensation) is called the Claisen-Schmidt condensation. This reaction is named after two of its pioneering investigators Rainer Ludwig Claisen and J. G. Schmidt, who independently published on this topic in 1880 and 1881. An example is the synthesis of dibenzylideneacetone. Quantitative yields in Claisen-Schmidt reactions have been reported in the absence of solvent using sodium hydroxide as the base and plus benzaldehydes.

Mechanism

Base catalyzed aldol reaction (shown using ⁻OCH₃ as base)

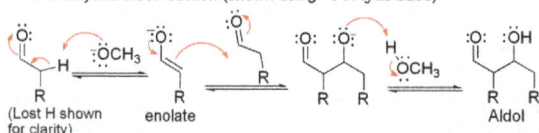

Acid catalyzed aldol reaction

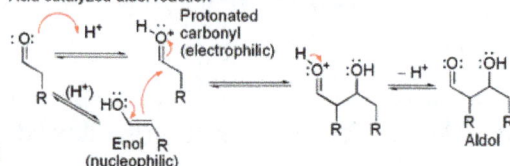

Base catalyzed dehydration (sometimes written as a single step)

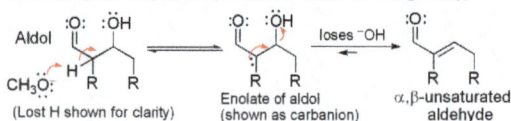

Acid catalyzed dehydration

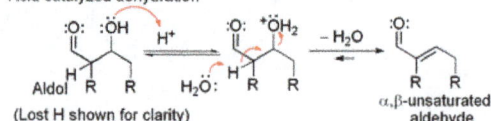

The first part of this reaction is an aldol reaction, the second part a dehydration—an elimination reaction (Involves removal of a water molecule or an alcohol molecule). Dehydration may be accompanied by decarboxylation when an activated carboxyl group is present. The aldol addition product can be dehydrated via two mechanisms; a strong base like potassium *t*-butoxide, potassium hydroxide or sodium hydride in an enolate mechanism, or in an acid-catalyzed enol mechanism. Depending on the nature of the desired product, the aldol condensation may be carried out under two broad types of conditions: kinetic control or thermodynamic control.

Condensation Types

It is important to distinguish the aldol condensation from other addition reactions of carbonyl compounds.

- When the base is an amine and the active hydrogen compound is sufficiently activated the reaction is called a Knoevenagel condensation.

- In a Perkin reaction the aldehyde is aromatic and the enolate generated from an anhydride.

- A Claisen condensation involves two ester compounds.

- A Dieckmann condensation involves two ester groups in the *same molecule* and yields a cyclic molecule

- A Henry reaction involves an aldehyde and an aliphatic nitro compound.

- A Robinson annulation involves a α,β-unsaturated ketone and a carbonyl group, which first engage in a Michael reaction prior to the aldol condensation.

- In the Guerbet reaction, an aldehyde, formed *in situ* from an alcohol, self-condenses to the dimerized alcohol.

- In the Japp–Maitland condensation water is removed not by an elimination reaction but by a nucleophilic displacement

Aldox Process

In industry the Aldox process developed by Royal Dutch Shell and Exxon, converts propylene and syngas directly to 2-ethylhexanol via hydroformylation to butyraldehyde, aldol condensation to 2-ethylhexenal and finally hydrogenation.

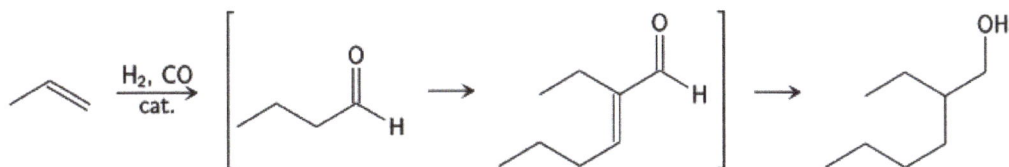

In one study crotonaldehyde is directly converted to 2-ethylhexanal in a palladium / Amberlyst / supercritical carbon dioxide system:

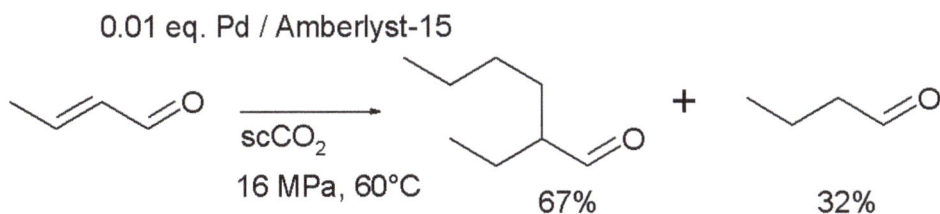

Scope

Ethyl 2-methylacetoacetate and campholenic aldehyde react in an Aldol condensation. The synthetic procedure is typical for this type of reaction. In the process, in addition to water, an equivalent of ethanol and carbon dioxide are lost in decarboxylation.

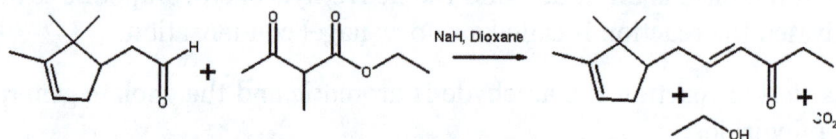

Ethyl glyoxylate 2 and diethyl 2-methylglutaconate 1 react to *isoprenetricarboxylic acid* 3 (isoprene skeleton) with sodium ethoxide. This reaction product is very unstable with initial loss of carbon dioxide and followed by many secondary reactions. This is believed to be due to steric strain resulting from the methyl group and the carboxylic group in the *cis*-dienoid structure.

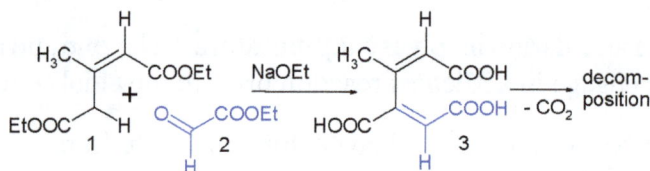

Occasionally an aldol condensation is buried in a multistep reaction or in catalytic cycle such as the one sketched below:

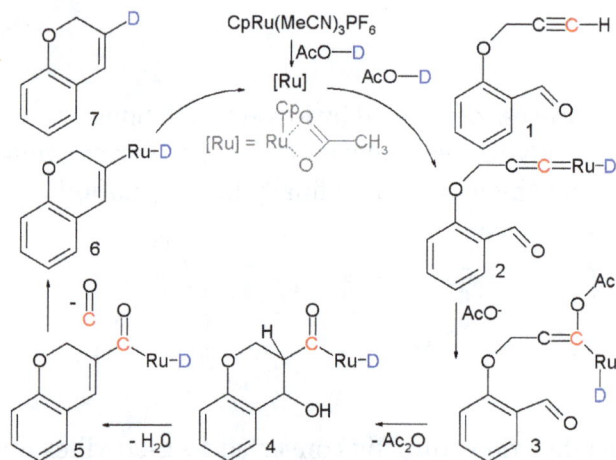

In this reaction an *alkynal* 1 is converted into a cycloalkene 7 with a ruthenium catalyst and the actual condensation takes place with intermediate 3 through 5. Support for the reaction mechanism is based on isotope labeling.

The reaction between menthone and anisaldehyde is complicated due to steric shielding of the ketone group. This obstacle is overcome by using a strong base such as potassium hydroxide and a very polar solvent such as DMSO in the reaction below:

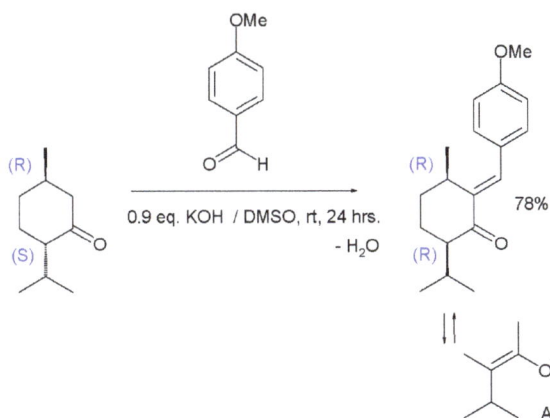

Due to epimerization through a common enolate ion (intermediate A) the reaction product has (R,R)-cis-configuration and not (R,S)-trans-configuration as in the starting material. Because it is only the cis isomer that precipitates from solution, this product is formed exclusively.

The reaction has become one of the most important methods for carbon-carbon bond formation. It consists of the reaction between two molecules of aldehydes or ketones that may be same or different. One of the reactants is converted into a nucleophile by forming its enolate in the presence of base and the second acts as an electrophile.

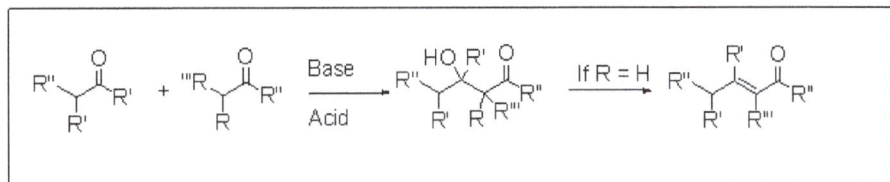

The geometry ((Z)- or (E)) of the enolate depends on the reaction conditions and the nature of the substituents. Strong base (e.g. LDA), low temperature and short reaction time lead to kinetic enolate , while weak base (e.g. hydroxide ion), high temperature and longer reaction time favour the formation of thermodynamic enolate.

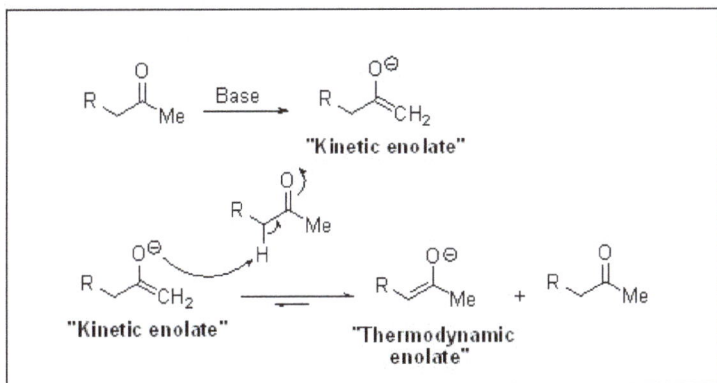

If the reactants are not the same, they can lead to the formation of diastereoisomers and their distribution depends on the reaction conditions and the nature of the substituents.

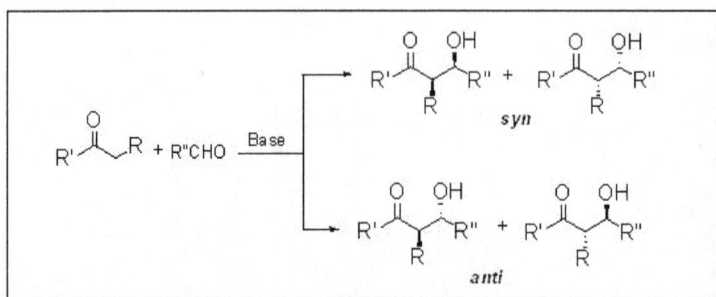

Under thermodynamic control, the (Z)- and (E)-forms of the enolates are in rapid equilibrium, and the product distribution is determined by the relative stabilities of the six-membered chair-shaped cyclic transition states that includes the metal counter-ion. Transition sate that leads to the syn product has R in the less stable axial position, whereas in that leading to anti product both R and R' are in the more stable equatorial position. The latter is therefore of lower energy, leading to a major anti product.

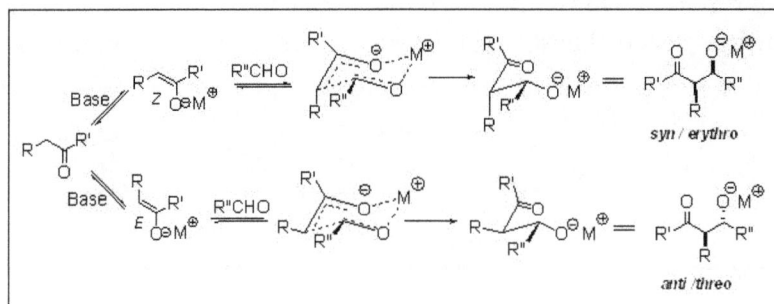

In contrast, under kinetic control, the (Z) and (E) enolates are formed rapidly and ir-reversibly, and their relative amounts determine the product distribution. For examples, for ketone $CH_3 CH_3 CO$ t -Bu, the (Z)-enolate is normally formed to afford the syn dias-teroisomer as a major product. But, the selectivity falls to 4:1 (syn : anti) when the size of the R is reduced from t -Bu to isopropyl. This is presumably because of the difference in steric repulsion of the methyl group with t-butyl and isopropyl groups.

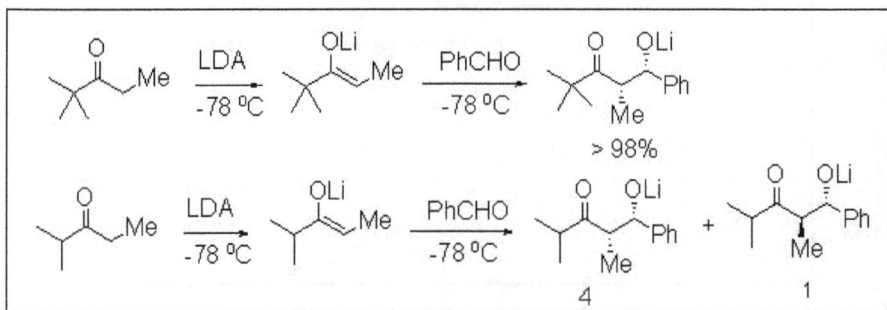

However, there is a general technique to increase the degree of diastereoselectivity. The enolate can be converted into silyl enol ethers that can be separated by distillation. The separated silyl enol ethers can then be converted into pure (Z)- or (E)-enoate by treatment with fluoride ion.

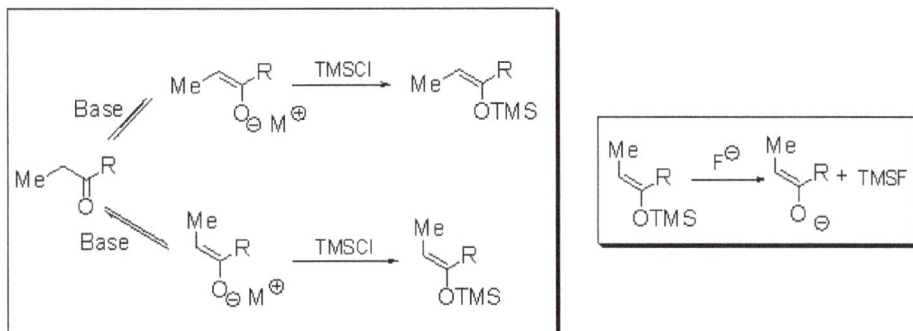

Asymmetric version of this reaction has also been well explored. For examples, chiral auxiliaries and chiral catalysts have been used as chiral source for asymmetric aldol reactions.

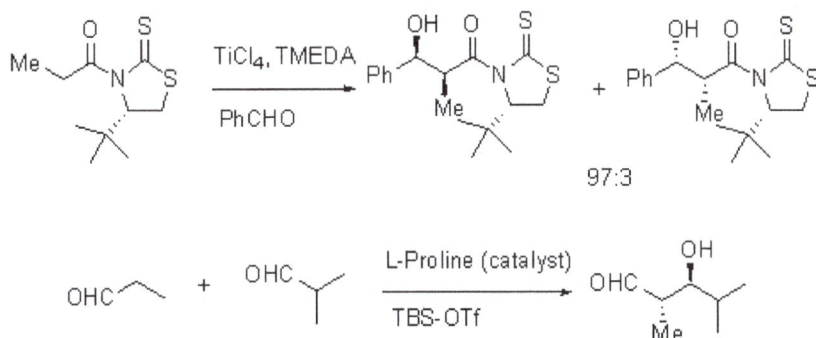

Examples

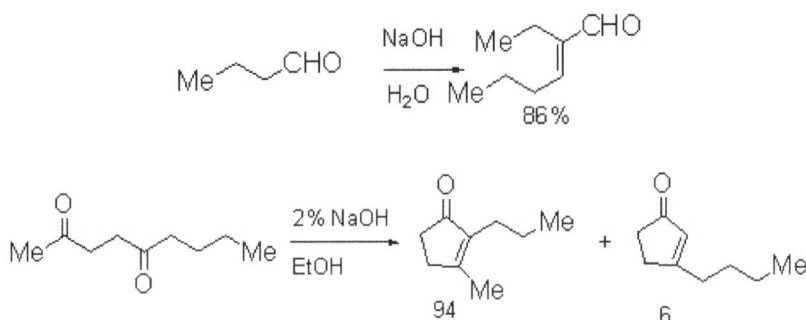

Reformatsky Reaction

The Reformatsky reaction (sometimes spelled Reformatskii reaction) is an organic reaction which condenses aldehydes or ketones, with α-halo esters, using a metallic zinc to form β-hydroxy-esters:

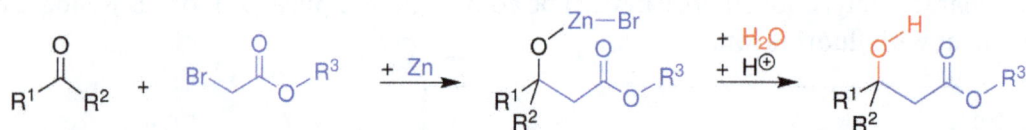

The organozinc reagent, also called a 'Reformatsky enolate', is prepared by treating an alpha-halo ester with zinc dust. Reformatsky enolates are less reactive than lithium enolates or clemison reagents and hence nucleophilic addition to the ester group does not occur. The reaction was discovered by Sergey Nikolaevich Reformatsky.

Structure of the Reagent

The crystal structures of the THF complexes of the Reformatsky reagents *tert*-butyl bromozincacetate and ethyl bromozincacetate have been determined. Both form cyclic eight-membered dimers in the solid state, but differ in stereochemistry: the eight-membered ring in the ethyl derivative adopts a tub-shaped conformation and has *cis* bromo groups and *cis* THF ligands, whereas in the *tert*-butyl derivative, the ring is in a chair form and the bromo groups and THF ligands are *trans*.

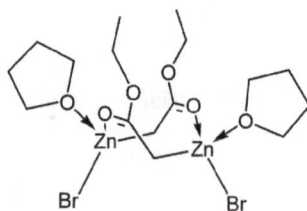

ethyl bromozincacetate dimer

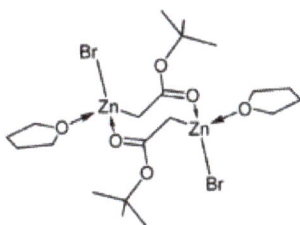

tert-butyl bromozincacetate dimer

Reaction Mechanism

Zinc metal is inserted into the carbon-halogen bond of the α-haloester by oxidative addition 1. This compound dimerizes and rearranges to form two zinc enolates 2. The oxygen on an aldehyde or ketone coordinates to the zinc to form the six-member chair like transition state 3. A rearrangement occurs in which zinc switches to the aldehyde or ketone oxygen and a carbon-carbon bond is formed 4. Acid workup 5,6 removes zinc to yield zinc(II) salts and a β-hydroxy-ester 7.

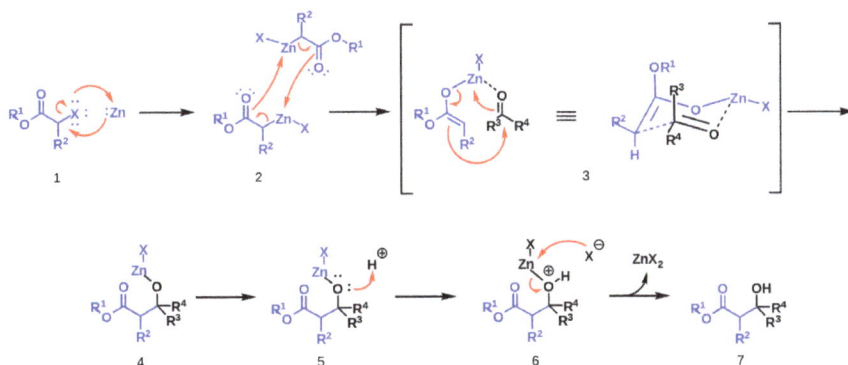

Variations

In one variation of the Reformatsky reaction an iodolactam is coupled with an aldehyde with triethylborane in toluene at -78 °C.

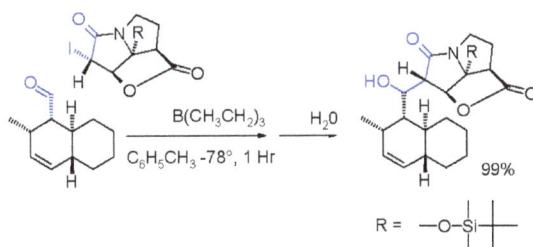

The Reformatsky Reaction

The enolate generated from an a -bromo ester with zinc reacts with an aldehyde or ketone to give an aldol-type product in diethyl ether.

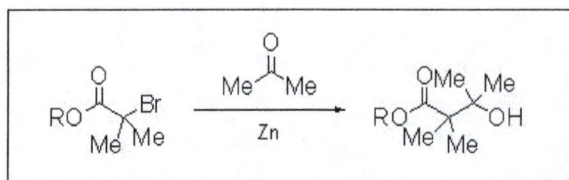

Mechanism

(Other metals can also be used)

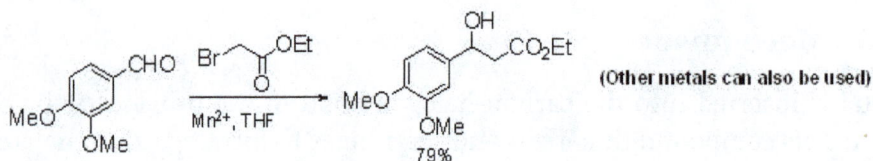

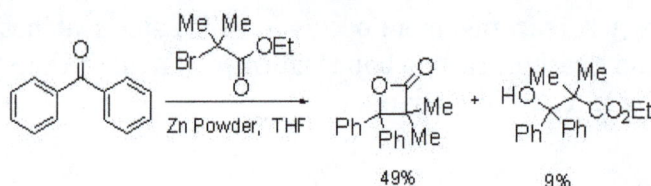

The Perkin Reaction

This process consists of the condensation of an acid anhydride with an aromatic alde-hyde using carboxylate ion).

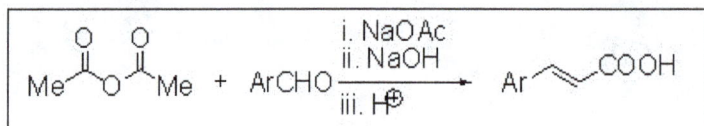

Mechanism

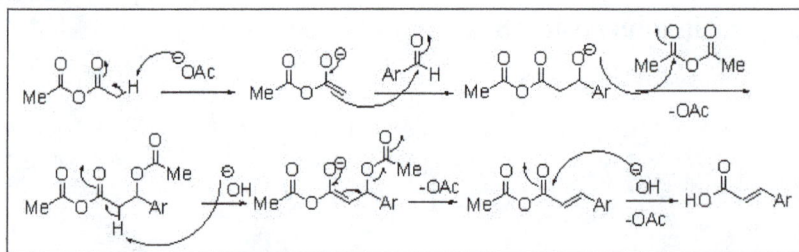

The Stobbe Condensation

The Stobbe condensation leads to attachment of three carbon chain to a ketonic carbon atom.

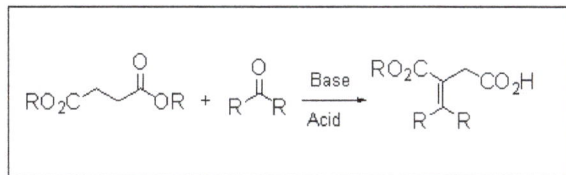

Mechanism

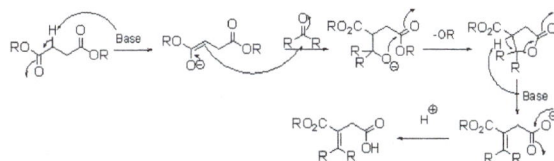

Examples

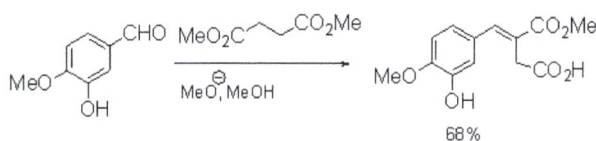

68%

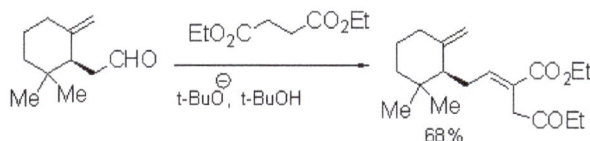

68%

The Darzen Reaction

The base-catalyzed condensation between an a -halo ester and an aldehyde or ketone affords glycidic ester.

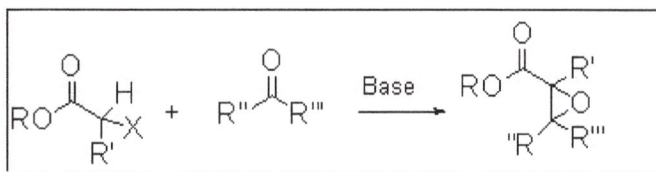

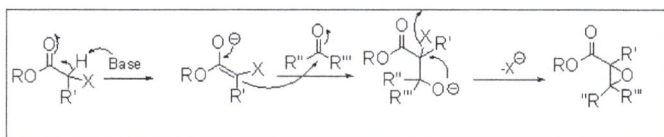

Examples

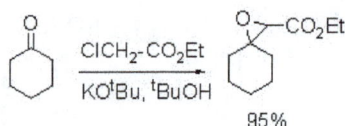

95%

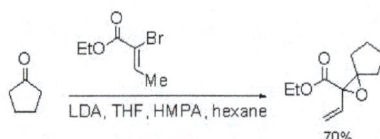

70%

The Knoevenagel Reaction

The condensation of methylene group bonded with two electron withdrawing groups with aldehydes or ketones using weak base is known as the Knoevenagel reaction. The reaction is more useful with aromatic than with aliphatic aldehydes.

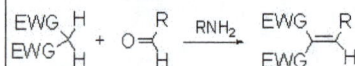

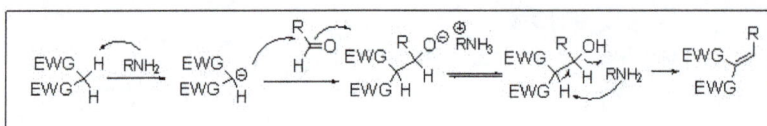

Examples

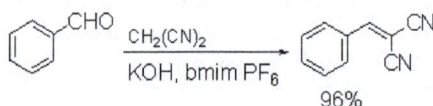

96%

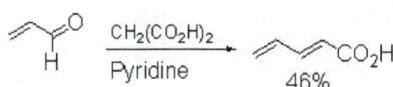

46%

Knoevenagel Condensation

The Knoevenagel condensation reaction is an organic reaction named after Emil Knoevenagel. It is a modification of the aldol condensation.

A Knoevenagel condensation is a nucleophilic addition of an active hydrogen compound to a carbonyl group followed by a dehydration reaction in which a molecule

pound to a carbonyl group followed by a dehydration reaction in which a molecule of water is eliminated (hence *condensation*). The product is often an αβ-unsaturated ketone (a conjugated enone).

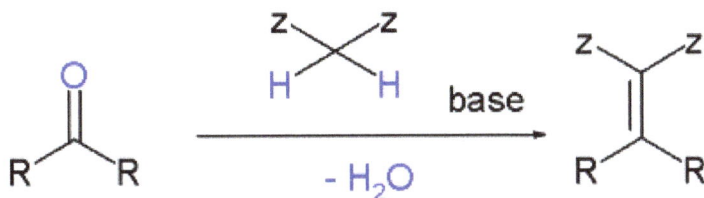

In this reaction the carbonyl group is an aldehyde or a ketone. The catalyst is usually a weakly basic amine. The active hydrogen component has the form

- Z–CH$_2$-Z or Z–CHR–Z for instance diethyl malonate, Meldrum's acid, ethyl acetoacetate or malonic acid, or cyanoacetic acid.

- Z–CHR$_1$R$_2$ for instance nitromethane.

where Z is an electron withdrawing functional group. Z must be powerful enough to facilitate deprotonation to the enolate ion even with a mild base. Using a strong base in this reaction would induce self-condensation of the aldehyde or ketone.

The Hantzsch pyridine synthesis, the Gewald reaction and the Feist–Benary furan synthesis all contain a Knoevenagel reaction step. The reaction also led to the discovery of CS gas.

Doebner Modification

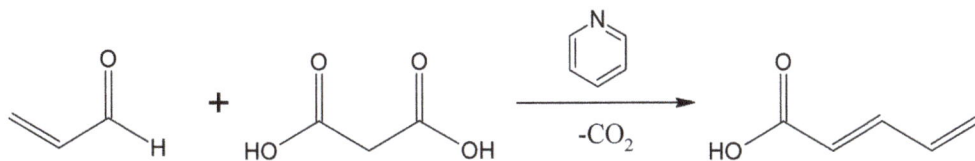

The Doebner modification of the Knoevenagel condensation. Acrolein and malonic acid react in pyridine to give trans-2,4-pentadienoic acid with the loss of carbon dioxide.

With malonic compounds the reaction product can lose a molecule of carbon dioxide in a subsequent step. In the so-called Doebner modification the base is pyridine. For example, the reaction product of acrolein and malonic acid in pyridine is *trans-2,4-Pentadienoic acid* with one carboxylic acid group and not two.

Scope

A Knoevenagel condensation is demonstrated in the reaction of 2-methoxybenzaldehyde 1 with the thiobarbituric acid 2 in ethanol using piperidine as a base. The resulting enone 3 is a charge transfer complex molecule.

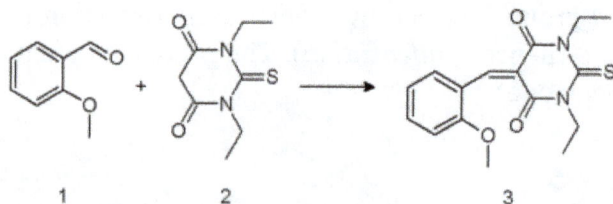

The Knoevenagel condensation is a key step in the commercial production of the anti-malarial drug lumefantrine (a component of Coartem):

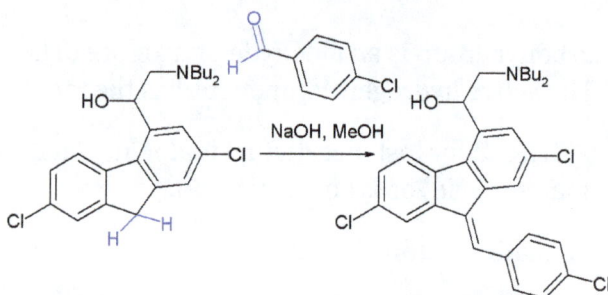

The initial reaction product is a 50:50 mixture of E and Z isomers but because both isomers equilibrate rapidly around their common hydroxyl precursor, the more stable Z-isomer can eventually be obtained.

A multicomponent reaction featuring a Knoevenagel condensation is demonstrated in this MORE synthesis with cyclohexanone, malononitrile and 3-amino-1,2,4-triazole:

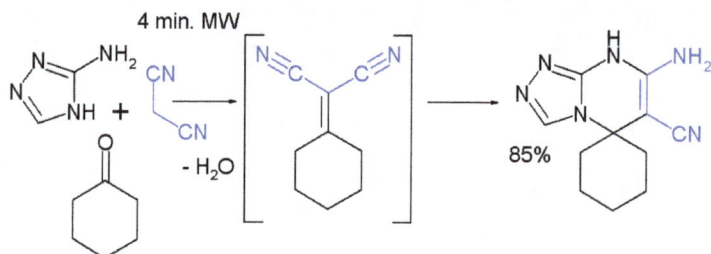

Weiss–Cook Reaction

The Weiss–Cook reaction consists in the synthesis of cis-bicyclo[3.3.0]octane-3,7-di-one employing an acetonedicarboxylic acid ester and a diacyl (1,2 ketone). The mechanism operates in same way as the Knoevenagel condensation:

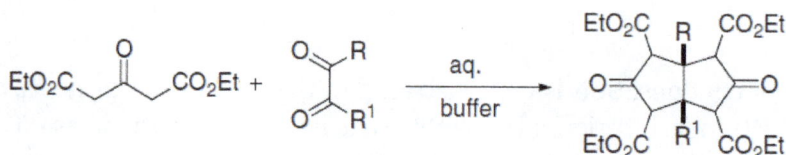

Darzens Reaction

The Darzens reaction (also known as the Darzens condensation or glycidic ester condensation) is the chemical reaction of a ketone or aldehyde with an α-haloester in the presence of base to form an α,β-epoxy ester, also called a "glycidic ester". This reaction was discovered by the organic chemist Auguste George Darzens in 1904.

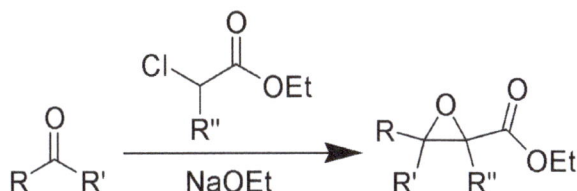

Reaction Mechanism

The reaction process begins when a strong base is used to form a carbanion at the halogenated position. Because of the ester, this carbanion is a resonance-stabilized enolate, which makes it relatively easy to form. This nucleophilic structure attacks another carbonyl component, forming a new carbon–carbon bond. These first two steps are similar to a base-catalyzed aldol reaction. The oxygen anion in this aldol-like product then does an intramolecular S_N2 attack on the formerly-nucleophilic halide-bearing position, displacing the halide to form an epoxide. This reaction sequence is thus a condensation reaction, since there is a net loss of HCl when the two reactant molecules join.

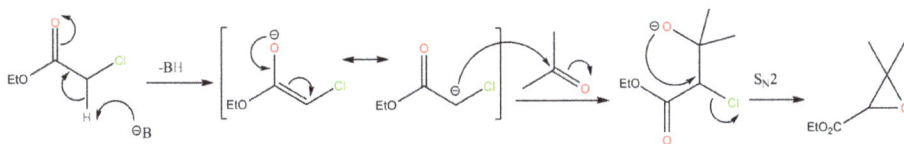

The primary role of the ester is to enable the initial deprotonation to occur, and other carbonyl functional groups can be used instead. If the starting material is an α-halo amide, the product is an α,β-epoxy amide. If an α-halo ketone is used, the product is an α,β-epoxy ketone.

Any sufficiently strong base can be used for the initial deprotonation. However, if the starting material is an ester, the alkoxide corresponding to the ester side-chain is commonly in order to prevent complications due to potential acyl exchange side reactions.

Stereochemistry

Depending on the specific structures involved, the epoxide may exist in *cis* and *trans* forms. A specific reaction may give only *cis*, only *trans*, or a mixture of the two. The specific stereochemical outcome of the reaction is affected by several aspects of the intermediate steps in the sequence.

The initial stereochemistry of the reaction sequence is established in the step where the carbanion attacks the carbonyl. Two sp^3 (tetrahedral) carbons are created at this stage, which allows two different diastereomeric possibilities of the halohydrin intermediate. The most likely result is due to chemical kinetics: whichever product is easier and faster to form will be the major product of this reaction. The subsequent S_N2 reaction step proceeds with stereochemical inversion, so the *cis* or *trans* form of the epoxide is controlled by the kinetics of an intermediate step. Alternately, the halohydrin can epimerize due to the basic nature of the reaction conditions prior to the S_N2 reaction. In this case, the initially formed diastereomer can convert to a different one. This is an equilibrium process, so the *cis* or *trans* form of the epoxide is controlled by chemical thermodynamics—the product resulting from the more stable diastereomer, regardless of which one was the kinetic result.

Alternative Reactions

Glycidic esters can also be obtained via nucleophilic epoxidation of an α,β-unsaturated ester, but that approach requires synthesis of the alkene substrate first whereas the Darzens condensation allows formation of the carbon–carbon connectivity and epoxide ring in a single reaction.

Subsequent Reactions

The product of the Darzens reaction can be reacted further to form various types of compounds. Hydrolysis of the ester can lead to decarboxylation, which triggers a rearrangement of the epoxide into a carbonyl (4). Alternately, other epoxide rearrangements can be induced to form other structures.

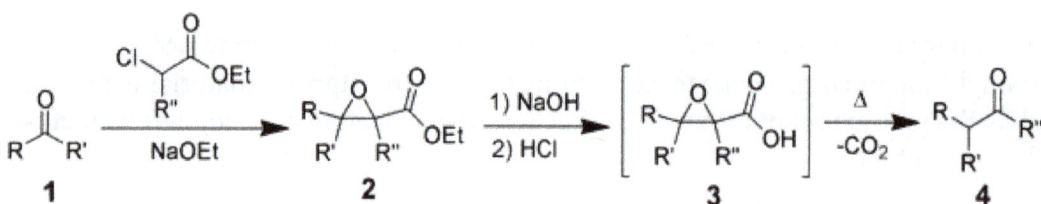

Perkin Reaction

The Perkin reaction is an organic reaction developed by William Henry Perkin that is used to make cinnamic acids. It gives an α,β-unsaturated aromatic acid by the aldol condensation of an aromatic aldehyde and an acid anhydride, in the presence of an alkali salt of the acid. The alkali salt acts as a base catalyst, and other bases can be used instead.

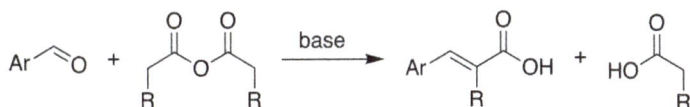

Reaction Mechanism

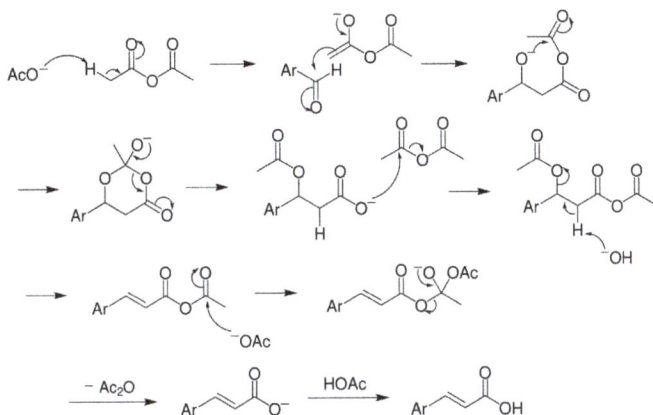

The above mechanism is not universally accepted, as several other versions exist, including decarboxylation without acetic group transfer.

Applications

- One notable application for the Perkin reaction is in the laboratory synthesis of the phytoestrogenic stilbene resveratrol (c.f. Fo-ti).

Claisen Condensation

The Claisen condensation is a carbon–carbon bond forming reaction that occurs between two esters or one ester and another carbonyl compound in the presence of a strong base, resulting in a β-keto ester or a β-diketone. It is named after Rainer Ludwig Claisen.

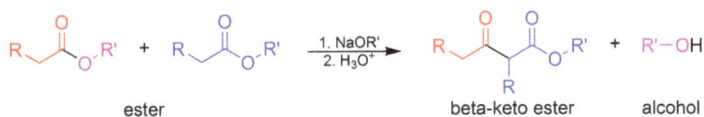

Requirements

At least one of the reagents must be enolizable (have an α-proton and be able to undergo deprotonation to form the enolate anion). There are a number of different combinations of enolizable and nonenolizable carbonyl compounds that form a few different types of Claisen condensations.

The base used must not interfere with the reaction by undergoing nucleophilic substitution or addition with a carbonyl carbon. For this reason, the conjugate sodium alkoxide base of the alcohol formed (e.g. sodium ethoxide if ethanol is formed) is often used, since the alkoxide is regenerated. In mixed Claisen condensations, a non-nucleophilic base such as lithium diisopropylamide, or LDA, may be used, since only one compound is enolizable. LDA is not commonly used in the classic Claisen or Dieckmann condensations due to enolization of the electrophilic ester.

The alkoxy portion of the ester must be a relatively good leaving group. Methyl and ethyl esters, which yields methoxide and ethoxide, respectively, are commonly used.

Types

- The classic Claisen condensation, a self-condensation between two molecules of a compound containing an enolizable ester.

ethyl acetate   ethyl acetoacetate   ethanol
                             (75%)

- The mixed (or "crossed") Claisen condensation, where one enolizable ester or ketone and one nonenolizable ester are used.

ethyl benzoate   acetophenone   1,3-diphenylpropane-1,3-dione   ethanol

- The Dieckmann condensation, where a molecule with two ester groups reacts intramolecularly, forming a cyclic β-keto ester. In this case, the ring formed must not be strained, usually a 5- or 6-membered chain or ring.

diethyl adipate   ethyl 2-oxocyclopentanecarboxylate   ethanol

Mechanism

In the first step of the mechanism, an α-proton is removed by a strong base, resulting in the formation of an enolate anion, which is made relatively stable by the delocalization of electrons. Next, the carbonyl carbon of the (other) ester is nucleophilically attacked by the enolate anion. The alkoxy group is then eliminated (resulting in (re)generation of the alkoxide), and the alkoxide removes the newly formed doubly α-proton to form a new, highly resonance-stabilized enolate anion. Aqueous acid (e.g. sulfuric acid or phosphoric acid) is added in the final step to neutralize the enolate and any base still present. The newly formed β-keto ester or β-diketone is then isolated. Note that the reaction requires a stoichiometric amount of base as the removal of the doubly α-proton thermodynamically drives the otherwise endergonic reaction. That is, Claisen condensation does not work with substrates having only one α-hydrogen because of the driving force effect of deprotonation of the β-keto ester in the last step.

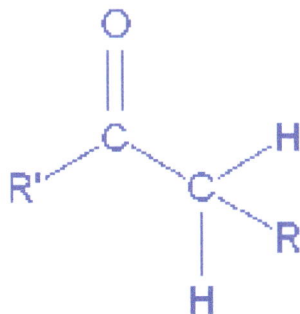

Stobbe Condensaion

The Stobbe condensation is a modification specific for the diethyl ester of succinic acid requiring less strong bases. An example is its reaction with benzophenone:

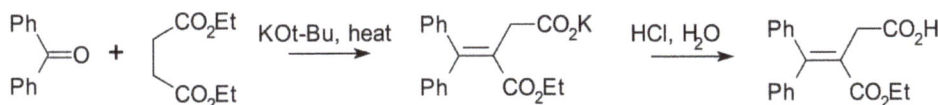

A reaction mechanism that explains the formation of both an ester group and a carboxylic acid group is centered on a lactone intermediate (5):

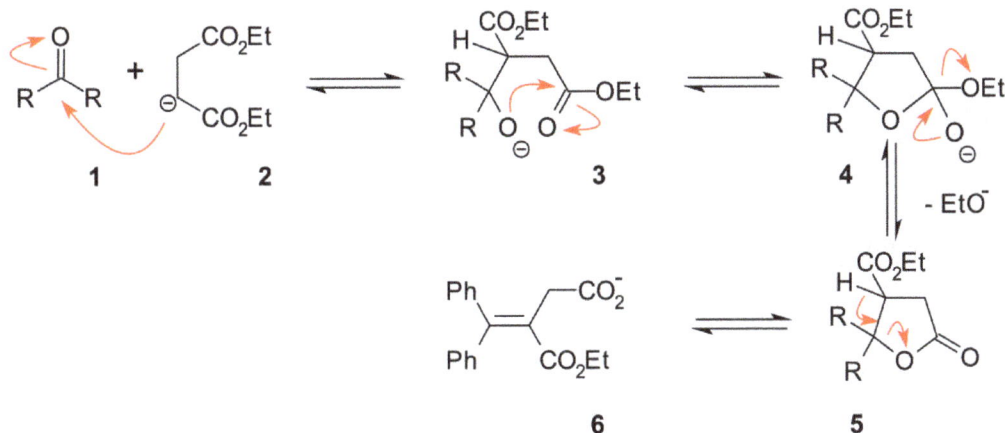

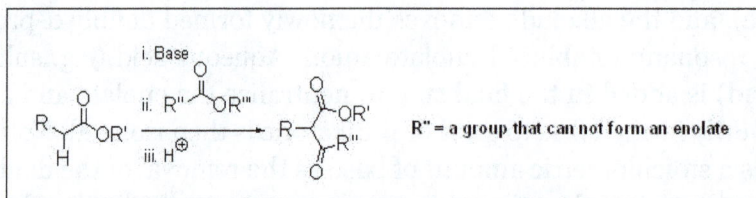

R" = a group that can not form an enolate

Mechanism

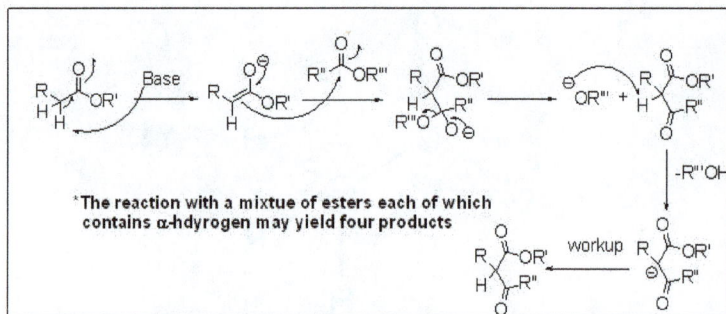

*The reaction with a mixtue of esters each of which contains α-hdyrogen may yield four products

Examples

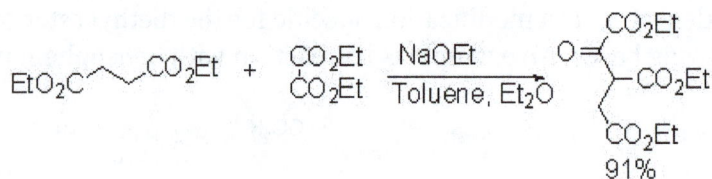

91%

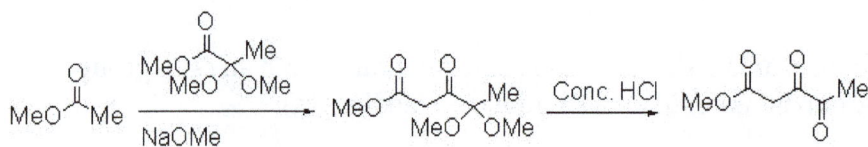

Dieckmann Condensation

The Dieckmann condensation is the intramolecular chemical reaction of diesters with base to give β-ketoesters. It is named after the German chemist Walter Dieckmann (1869–1925). The equivalent intermolecular reaction is the Claisen condensation.

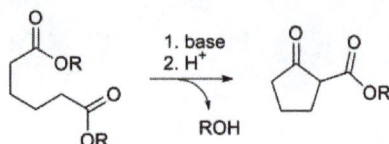

Reaction Mechanism

Deprotonation of an ester at the α-position generates an enolate ion which then undergoes a 5-exo-trig nucleophilic attack to give a cyclic enol. Protonation with a Brønsted-Lowry acid (H_3O^+ for example) re-forms the β-keto ester.

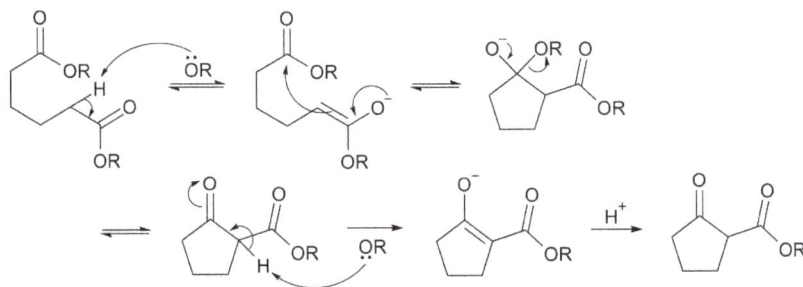

Owing to the steric stability of five- and six-membered ring structures, these will preferentially be formed. So 1,6 diesters will form five-membered cyclic β-keto esters, while 1,7 diesters will form six-membered β-keto esters.

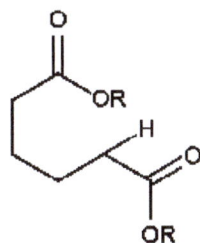

Reaction of mechanism

Condensation of the diesters of having C_6 and C_7 can be accomplished to afford five and six membered cyclic β-ketoesters. The diesters of short-chain do not show cyclization, while diesters with C_8 and C_9 provide the cyclized products in fewer yields.

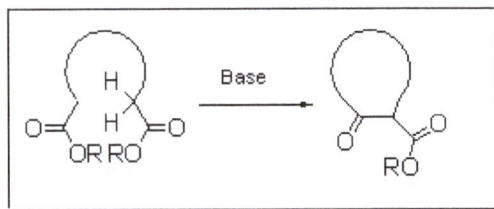

Mechanism

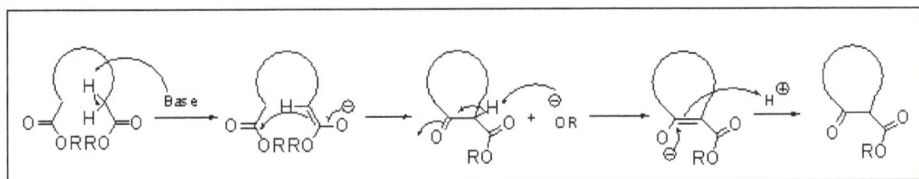

Examples

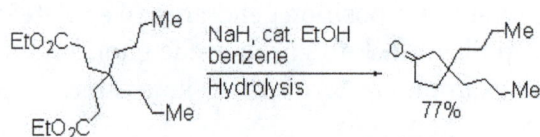

77%

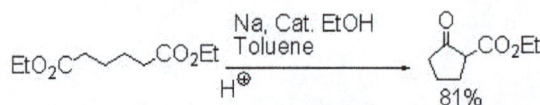

81%

Thorpe Reaction

The Thorpe reaction is a chemical reaction described as a self-condensation of aliphatic nitriles catalyzed by base to form enamines. The reaction was discovered by Jocelyn Field Thorpe.

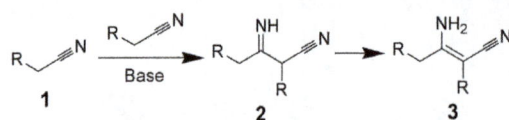

Thorpe–Ziegler Reaction

The Thorpe–Ziegler reaction (named after Jocelyn Field Thorpe and Karl Ziegler), or Ziegler method, is the intramolecular modification with a dinitrile as a reactant and a cyclic ketone as the final reaction product after acidic hydrolysis. The reaction is conceptually related to the Dieckmann condensation.

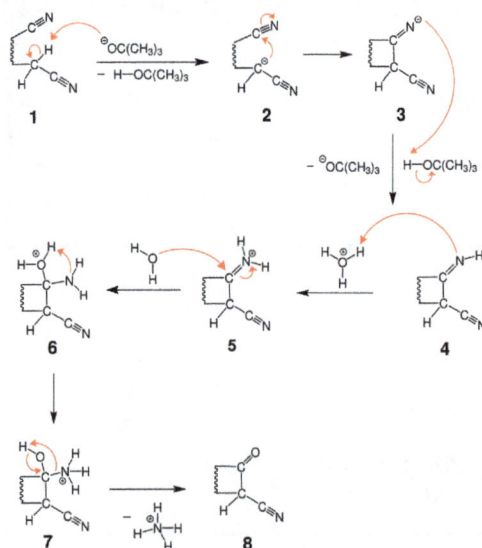

The cyclization of dinitriles using base can be accomplished. Although it is similar to the Dieckmann reaction, the former is often better compared to the latter.

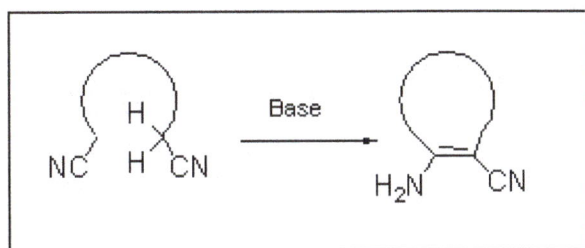

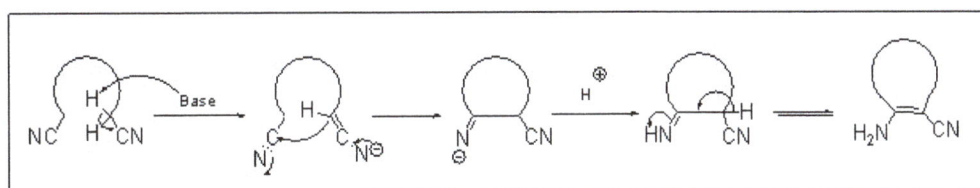

Examples

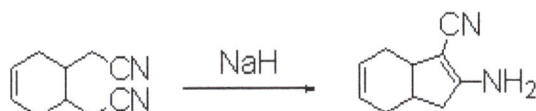

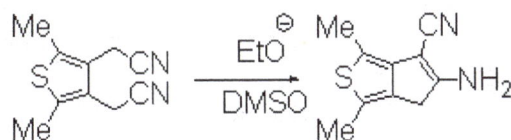

Enamine

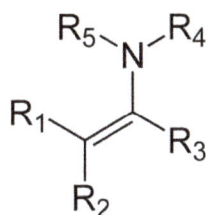

The general structure of an enamine

An enamine is an unsaturated compound derived by the condensation of an aldehyde or ketone with a secondary amine. Enamines are versatile intermediates.

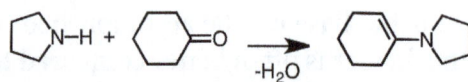

Condensation to give an enamine.

The word "enamine" is derived from the affix *en-*, used as the suffix of alkene, and the root *amine*. This can be compared with enol, which is a functional group containing both alkene (*en-*) and alcohol (*-ol*). Enamines are considered to be nitrogen analogs of enols.

If one of the nitrogen substituents is a hydrogen atom, H, it is the tautomeric form of an imine. This usually will rearrange to the imine; however there are several exceptions (such as aniline). The enamine-imine tautomerism may be considered analogous to the keto-enol tautomerism. In both cases, a hydrogen atom switches its location between the heteroatom (oxygen or nitrogen) and the second carbon atom.

Enamines are both good nucleophiles and good bases. Their behavior as carbon-based nucleophiles is explained with reference to the following resonance structures.

Reactions

Formation

Enamines are labile and therefore chemically useful moieties which can be easily produced from commercially available starting reagents. A common route for enamine production is via an acid-catalyzed nucleophilic reaction of ketone (Stork, 1963) or aldehyde (Mannich/Davidsen 1936) species containing an α-hydrogen with secondary amines. Acid catalysis is not always required, if the pKa of the reacting amine is sufficiently high (for example, pyrrolidine- pKa 11.26). If the pKa of the reacting amine is low, however, then acid catalysis is required through both the addition and the dehydration steps (common dehydrating agents include $MgSO_4$ and Na_2SO_4). Primary amines are usually not used for enamine synthesis due to the preferential formation of the more thermodynamically stable imine species. Methyl ketone self-condensation is a side-reaction which can be avoided through the addition of $TiCl_4$ into the reaction mixture (to act as a water scavenger). An example of an aldehyde reacting with a secondary amine to form an enamine via a carbinolamine intermediate is shown below:

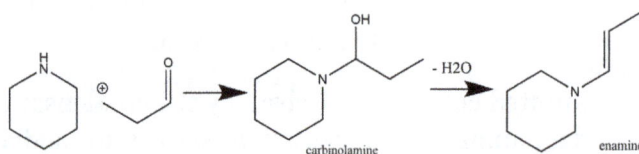

Alkylation

Even though enamines are more nucleophilic than their enol counterparts, they can still react selectively, rendering them particularly useful for alkylation reactions. The enamine nucleophile can attack haloalkanes to form the alkylated iminium salt intermediate which then hydrolyzes to regenerate a ketone (a starting material in enamine synthesis). This reaction was pioneered by Gilbert Stork, and is sometimes referred to by the name of its inventor. Analogously, this reaction can be used as an effective means of acylation. A variety of alkylating and acylating agents including benzylic, allylic halides can be used in this reaction.

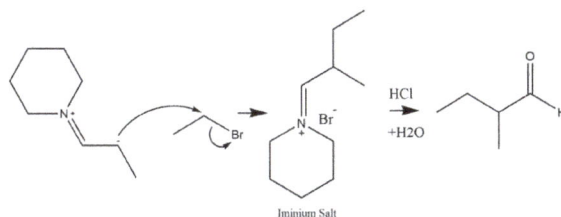

Iminium Salt

Acylation

In a reaction much similar to the enamine alkylation, enamines can be acylated to form a final dicarbonyl product. The enamine starting material undergoes a nucleophilic addition to acyl halides forming the iminium salt intermediate which can hydrolyze in the presence of acid.

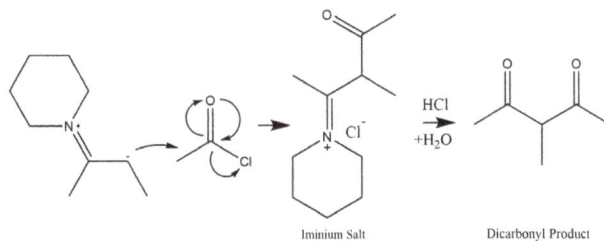

Iminium Salt   Dicarbonyl Product

Metalloenamines

Strong bases such as (LiNR2 can be used to deprotonate imines and form metalloenamines. Metalloenamines can prove synthetically useful due to their nucleophilicity (they are more nucleophilic than enolates). Thus they are better able to react with weaker electrophiles (for example, they can be used to open epoxides.) Most prominently, these reactions have allowed for asymmetric alkylations of ketones through transformation to chiral intermediate metalloenamines.

Halogenation

β-halo immonium compounds can be synthesized through the reaction of enamines with halides in diethyl ether solvent. Hydrolysis will result in the formation of α-halo ketones. Chlorination, bromination, and even iodination have been shown to be possible. The general reaction is shown below:

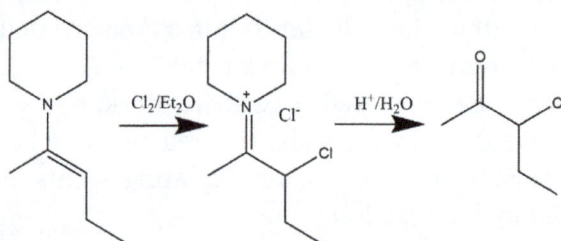

Oxidative Coupling

Enamines can be efficiently cross-coupled with enol silanes through treatment with Ce(IV) ammonium nitrate. These reactions were reported by the Narasaka group in 1935, providing a route to stable enamines as well as one instance of a 1,4 diketone (derived from a morpholine amine reagent). Later, these results were exploited by the McMillan group with the development of an organocatalyst which used the Narasaka substrates to produce 1,4 dicarbonyls enantioselectively, with good yields. Oxidative dimerization of aldehydes in the presence of amines proceeds through the formation of an enamine followed by a final pyrrole formation. This method for symmetric pyrrole synthesis was developed in 2010 by the Jia group, as a valuable new pathway for the synthesis of pyrrole-containing natural products.

Hajos–Parrish–Eder–Sauer–Wiechert Reaction

This reaction, reported in 1971 by several research teams, is an early example of an enantioselective catalytic reaction in organic chemistry. Its scope has been modified and expanded through the development of related reactions including the Michael addition, asymmetric aldol reaction, and the Mannich reaction. This reaction has likewise been used to perform asymmetric Robinson annulations. The general scheme of this reaction follows:

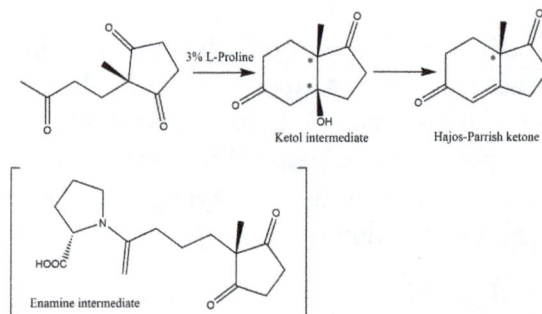

This is an example of a 6-enolendo aldolization.

A prominent example of proline catalysis is the addition of acetone or hydroxyacetone to a diverse set of aldehydes catalyzed by 20-30% proline catalyst loading with high (>99%) enantioselectivity yielding diol products. This chemistry was initially studied by the Barbas group at the Scripps Research Institute, and later refined by List and Notz who used the aforementioned reaction to produce diol products as follows:

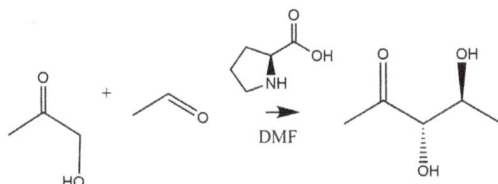

Proline Catalyzed Enolexo Aldolizations

These reactions are a recent development in contrast to their enolendo counterparts. Dicarbonyl (dials,diketones) can be converted to anti-aldol products with a 10% L-proline catalyst loading. This is an example of an enolexo intramolecular aldolization.

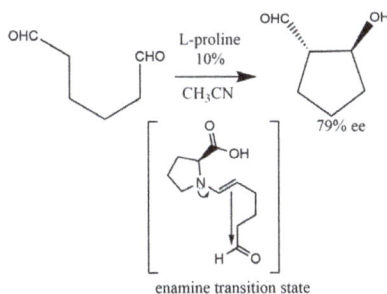

enamine transition state

Annulation

Enamines chemistry has been implemented for the purposes of producing a one-pot enantioselective version of the Robinson annulation. The Robinson annulation, published by Robert Robinson in 1935, is a base-catalyzed reaction that combines a ketone and a methyl vinyl ketone (commonly abbreviated to MVK) to form a cyclohexenone fused ring system. This reaction may be catalyzed by proline to proceed through chiral enamine intermediates which allow for good stereoselectivity. This is important, in particular in the field of natural product synthesis, for example, for the synthesis of the Wieland-Mescher ketone – a vital building block for more complex biologically active molecules.

Reactivity

Enamines act as nucleophiles that require less acid/base activation for reactivity than their enolate counterparts. They have also been shown to offer a greater selectivity with less side reactions. There is a gradient of reactivity among different enamine types, with a greater reactivity offered by ketone enamines than their aldehyde counterparts. Cyclic ketone enamines follow a reactivity trend where the five membered ring is the most reactive due to its maximally planar conformation at the nitrogen, following the trend 5>8>6>7 (the seven membered ring being the least reactive). This trend has been attributed to the amount of p-character on the nitrogen lone pair orbital- the higher p character corresponding to a greater nucleophilicity because the p-orbital would allow for donation into the alkene π- orbital. Analogously, if the N lone pair participates in stereoelectronic interactions on the amine moiety, the lone pair will pop out of the plane (will pyramidalize) and compromise donation into the adjacent π C-C bond.

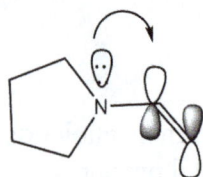

N lone pair donates effectively into the pi system   N lone pair cannot effectively interact with the pi system

There are many ways to modulate enamine reactivity in addition to altering the steric/electronics at the nitrogen center including changing temperature, solvent, amounts of other reagents, and type of electrophile. Tuning these parameters allows for the preferential formation of E/Z enamines and also affects the formation of the more/less substituted enamine from the ketone starting material.

Transition States

Proline Catalyzed Aldol Reactions

Proline-catalyzed aldol additions undergo a six-membered enamine transition state according to the Zimmerman-Traxler model. Addition of 20-30 mol% proline to acetone or hydroxyacetone catalyzes their addition to a diverse set of aldehydes with high (>99%) enantioselectivity yielding diol products. Proline and proline derivatives have been implemented as organocatalysts to promote asymmetric condensation reactions. An example of such a reaction proceeding through a six membered transition state is modelled as follows.

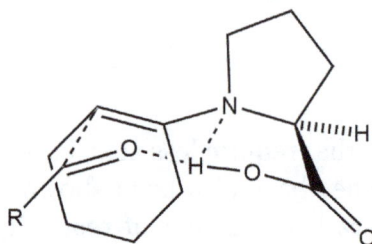

Intramolecular aldolization reactions that are catalyzed by proline likewise go through six-membered transition states- these transition states can enable the formation of either the enolexo or the enolendo product.

Enamines

The reaction of secondary amine with aldehyde or ketone that contains an a -hydrogen atom affords enamine. The process is driven to right by removing the water as it is formed, either by azotropic distillation or with molecular sieves.

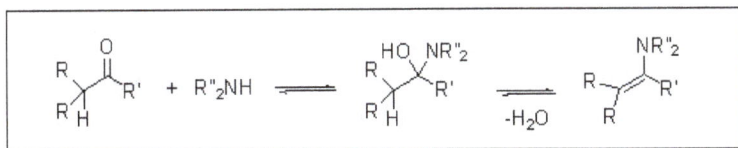

Similar to the enolates derived from ketones, enamines react with acid chlorides to give imine derivative that could be hydrolyzed to β -diketones.

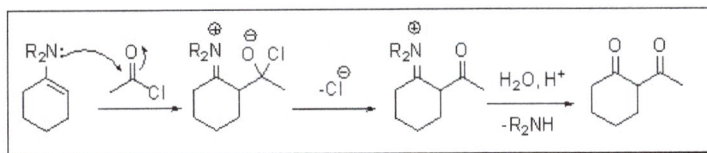

In case of unsymmetrical ketones, less substituted enanime forms as a major product.

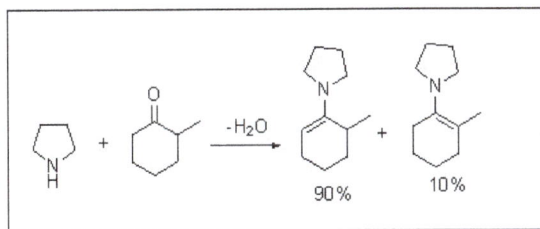

The Alkylation of Enolates

Enolates, like other nucleophiles, also undergo reaction with alkyl halides and sulfonates with the formation of carbon-carbon bonds. Depending on the reaction conditions and nature of the substrates, the reaction can occur either at oxygen atom or carbon atom of enolate.

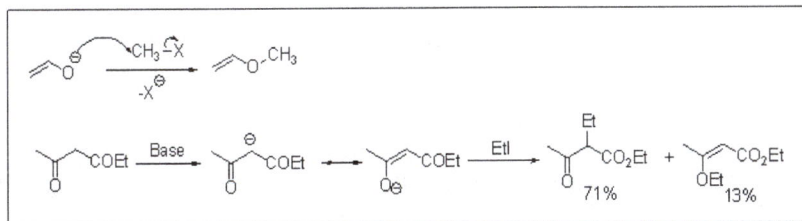

Alkylation of Monofunctional Compounds

Depends on the reaction conditions (kinetic vs thermodynamic control), enolate can be selectively alkylated.

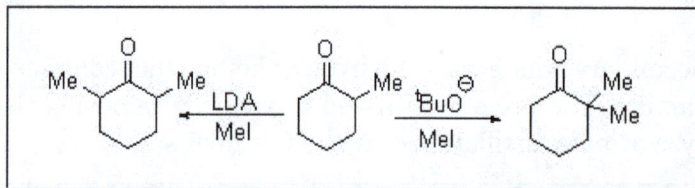

Kinetic control with LDA : Proton abstraction takes place at less hindered a -CH position and the reaction is faster and essentially irreversible.

Thermodynamic control with t BuOK : equilibration takes place between the two enolates and the methyl-substituted one, being the more stable, is present in high concentration.

However, when the highly substituted position is strongly sterically hindered, alkylation with t BuOK occurs at the less sterically substituted carbon.

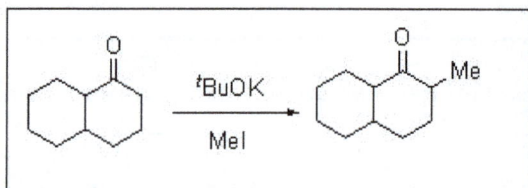

Alkylation of Bifunctional Compounds

A C-H bond adjacent to two electron withdrawing groups is more acidic than that adjacent to one electron withdrawing group and the alkylation could be carried out in milder conditions.

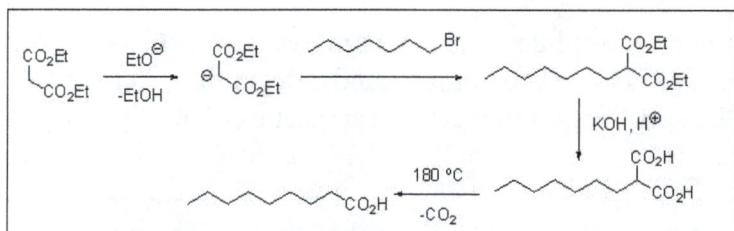

Addition of Enolates to Activated Alkenes

Enolates undergo addition to alkenes that are activated by conjugation to carbonyl, ester, nitro and nitril groups. These reactions are usually referred to as Michael addition.

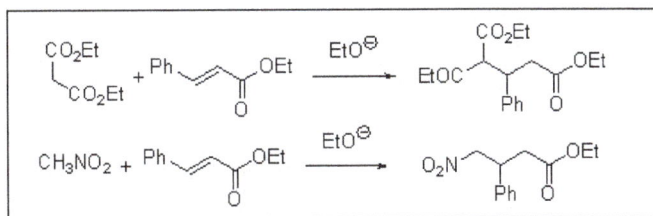

Examples

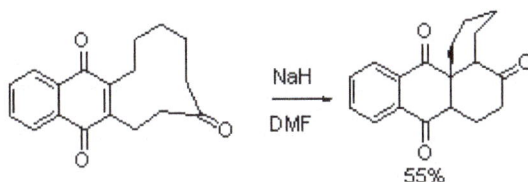

55%

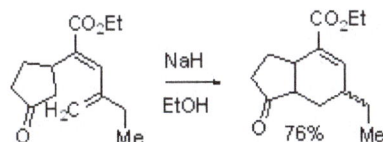

76%

Reactions Involving Alkynes

Acetylene and its monosubstituted derivatives are more acidic than alkenes and alkanes and take part in reactions with both carbonyl-containing compounds and alkyl halides in the presence of base.

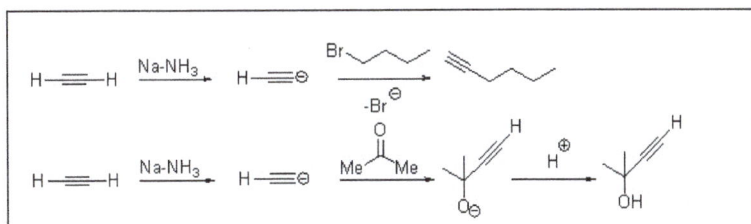

Application

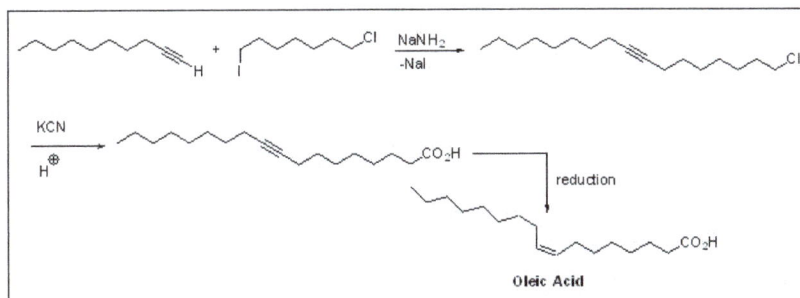

Oleic Acid

Reactions of Cyanides with Alkyl Halides and Sulfonates

HCN, like acetylene, is a weak acid whose anion may be generated by base and is reactive towards primary and secondary alkyl halides and sulfonates to give the corresponding nitriles. It is more convenient to introduce the cyanide as cyanide ion (e.g. NaCN, TMSCN) rather than as HCN. These reactions provide a way of extending aliphatic carbon chains by one carbon atom. Figure summarizes some of the useful transformations.

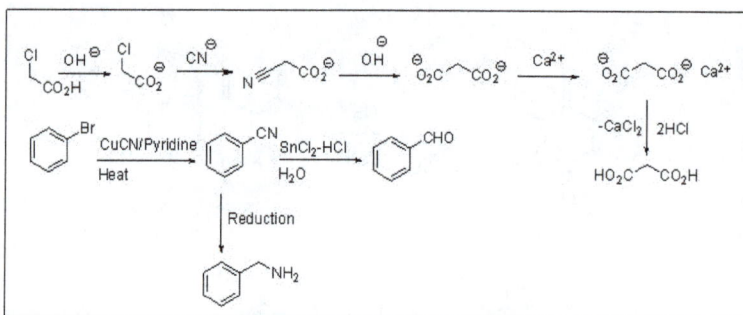

Mechanism for Stephen Aldehyde Synthesis (Stephen Reduction)

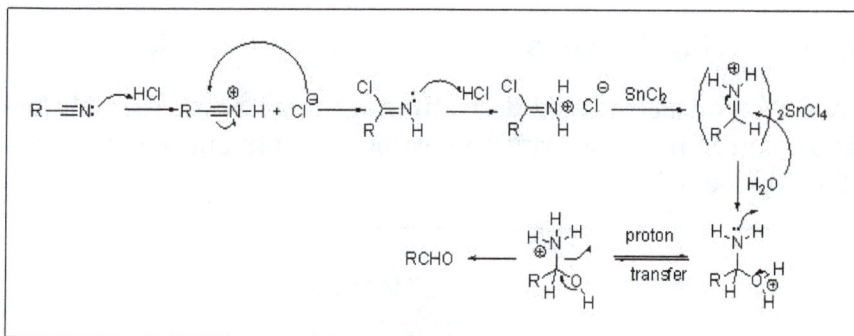

Reactions of Cyanides with Carbonyl Compounds

Cyanide ion adds to aldehydes and ketones to give cyanohydrins that can be hydrolyzed to a -hydroxy acids.

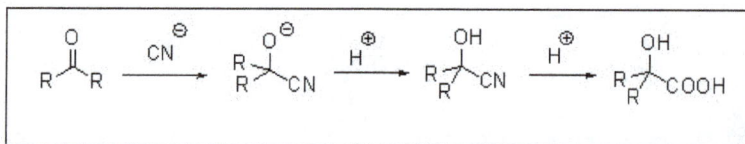

Asymmetric version of the process is also well explored. For example, chiral main chain polymer having Ti(VI) has been found to be an effective recyclable catalyst to obtain the cyanohydrin with up to 88% ee.

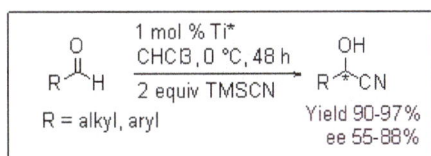
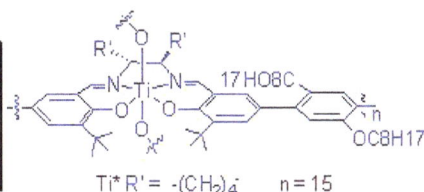

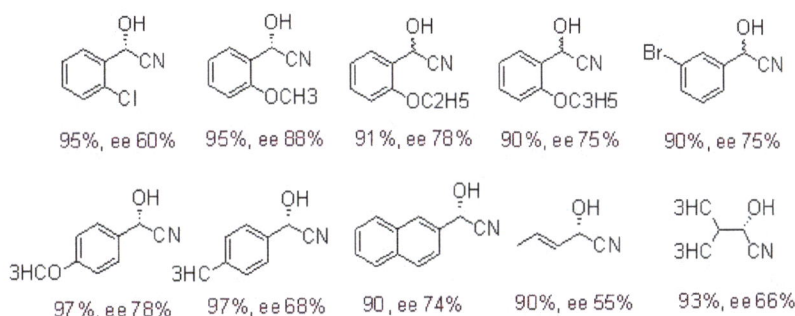

Imine that can be prepared from amine and aldehyde readily undergoes reaction with cyanide ion to give a -amino nitrile which can be hydrolyzed to a- amino acid (Strecker synthesis).

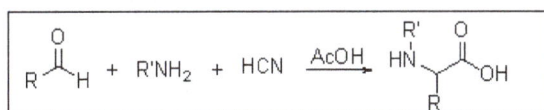

Mechanism

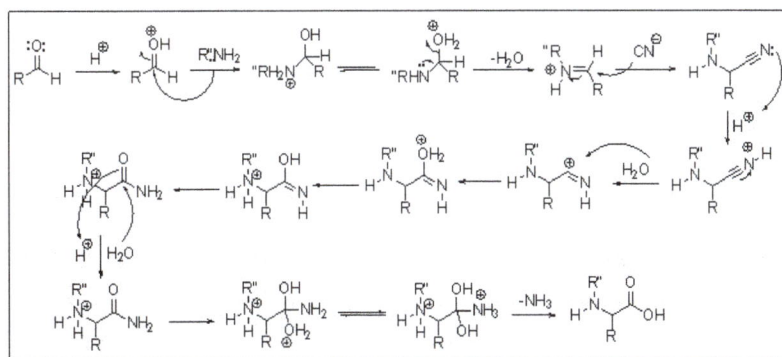

Examples

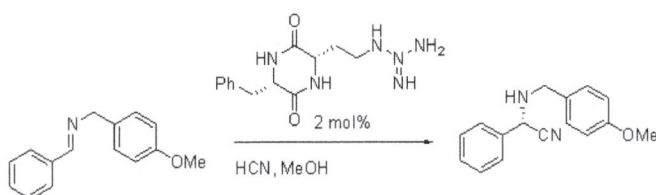

Acid Catalyzed Reactions

Principles

 of an electrophilic species with the aid of an acid which undergoes reaction with nucleophile. The electrophiles may be formed from an alkyl or acyl halide by treatment with a Lewis acid, e.g.

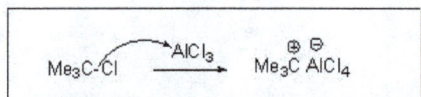

or, more generally, by the addition of a proton to a double bond.

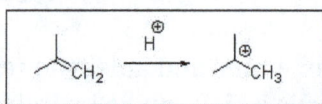

or a carbonyl group, as in the self-condensation of aldehydes and ketones, e.g.

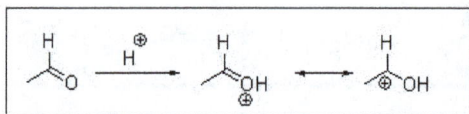

The electrophiles in the Mannich reaction are generated from an aldehydes and an amine in the presence of an acid e.g.

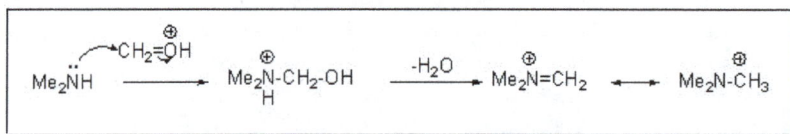

The nucleophiles may be an alkene, an enol or a compound of related type.

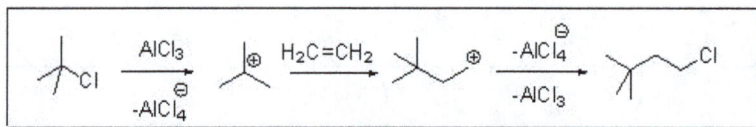

Self-Condensation of Alkenes

Alkenes undergo protonation with acid to give a carbocation that can add with second molecule of alkene to generate new carbocation which could eliminate a proton to give alkene. For an example, isobutylene with 60% sulfuric acid affords a 4:1 mixture of 2,4,4-trimethyl-1-pentene and 2,4,4-trimethyl-2-pentene. The reaction takes place by protonation of one molecule of the alkene to provide a carbocation that adds to the methylene group (Markonovkov's rule) of a second molecule. The new carbocation eliminates a proton to afford the products.

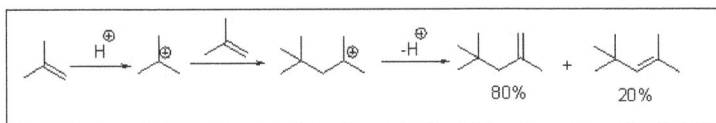

The reaction condition is crucial for the selectivity of the products. For an example, when dilute sulfuric acid is used, the first carbocation undergoes reaction preferentially with water to afford t-butanol as a major product.

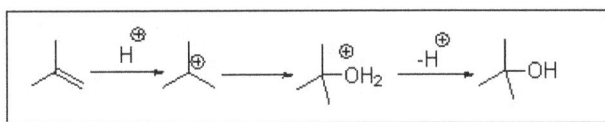

Alternatively, when more concentrated sulfuric acid is used, the second carbocation reacts with a further molecule of isobutylene to lead polymerization.

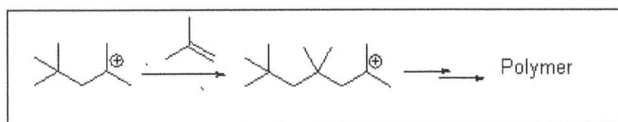

Furthermore, isooctane (used as high-octane fuel) is formed in the presence of isobutene. Under these conditions, the carbocation formed by dimerization of isobutylene abstracts hydride ion from isobutane. During the process a new t-butyl carbocation is generated which adds to isobutylene to lead a chain propagation.

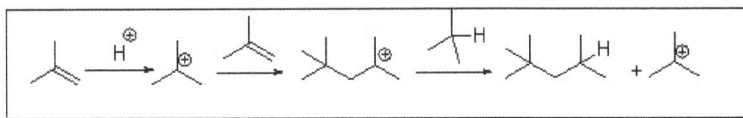

Dienes may undergo acid-catalyzed cyclizations provided the product is a stereochemically favored five or six membered ring. For example, in the presence of acid, ψ-ionone cyclizes to give a mixture of α - and β -ionones.

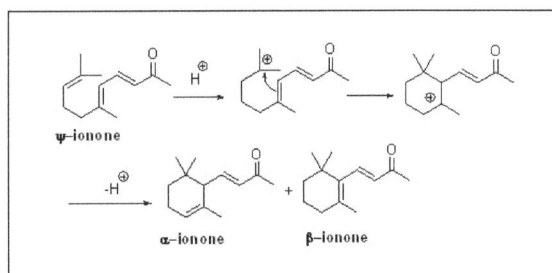

Reactions of Aldehydes and Ketones

The acid-catalyzed reactions of aldehydes and ketones can be broadly divided into four types: self-condensation, crossed-condensation, reactions of ketones with acid chlorides and acid anhydrides and reactions of 2-methyl pyridines and related compounds.

Self-Condensation

Aldehydes and ketones which are prone to undergo enolization proceed self-condensation in the presence of acids. The acid has dual functions: first, it enhances the reactivity of the carbonyl group towards the addition of a nucleophile, e.g.

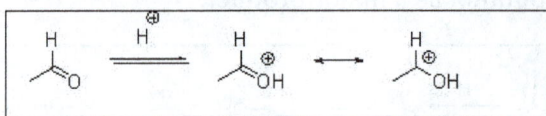

Next, it catalyzes the enol formation of the carbonyl compound, e.g.

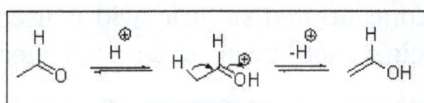

A molecule of enol then reacts with a molecule of the activated carbonyl compound.

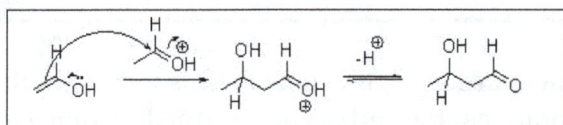

Dehydration via the enol generally follows:

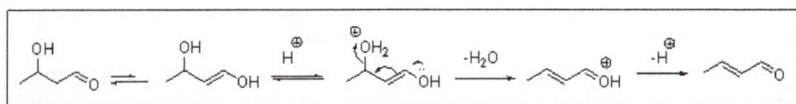

In case of ketones having α -hydrogen, further reaction possible. For an example, in presence of hydrogen chloride, acetone proceeds reaction to afford a mixture of mono- and di-condensation products.

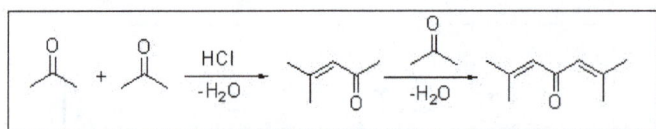

The product yield is usually low in acid-catalyzed reaction compared to that of the base-catalyzed reactions. However, in some instances the products are different, where the acid-catalyzed reactions are useful. For an example, the base-catalyzed reaction of benzaldehyde with butanone gives A as a main product whereas the acid-catalyzed reaction affords B as the major product.

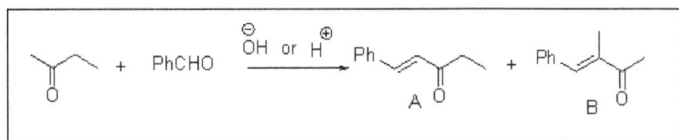

This is because, in the former, equilibria are established with the two possible aldol adducts and the main product is determined by the ease of dehydration to provide an aryl-conjugated product. Thus, the formation of A is favoured since the enolate C is generated faster compared to that of the more substitute enolate D.

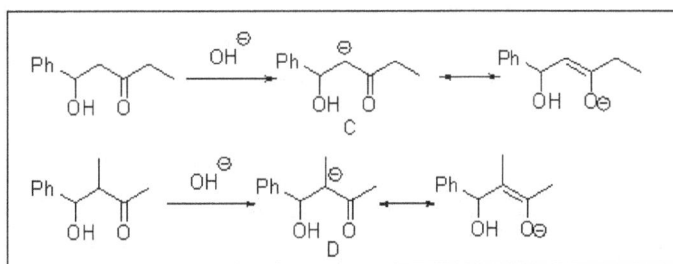

In contrast, in acid-catalyzed reaction the crucial step is not dehydration but the formation of the new C-C bond and the regioselectivity is governed by the rapid formation of the more substituted enol E compared to monoalkylated enol F. Thus, the formation of B is favoured compared to that of A.

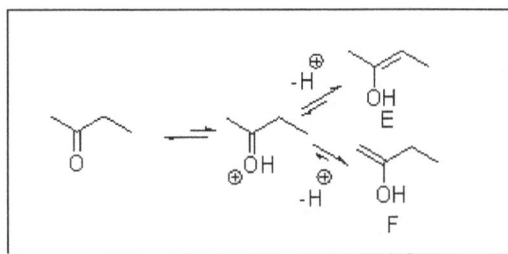

Finally, some aldehydes give cyclic polymers with acids. For example, acetal with trace of concentrated sulphuric acid provides the cyclic trimer, paraldehyde, and some of the cyclic tetramer, metaldehyde.

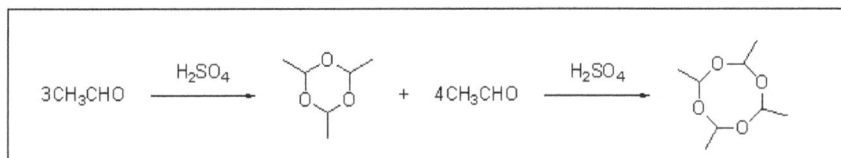

It is reversible and the aldehyde can be generated prior to use by warming with dilute acid. In case of ketones, acetone undergoes similar kind of polymerization with concentrated sulfuric acid producing 1,3,5-trimethylbenzene.

Crossed-Condensation

Similar to base-catalyzed reactions, crossed reactions between two enolizable carbonyl compounds may result in a mixture of four products. However, if only one of the two compounds can enolize and the other has the more reactive carbonyl group, a single product may then be formed. For example, the reaction of enolizable acetophenone with non-enolizable more reactive benzaldehyde can give single product in good yield.

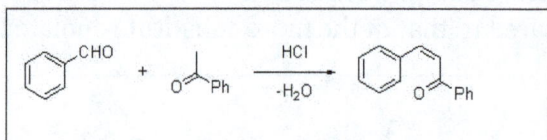

Reactions Between Ketones and Acid Chlorides or Anhydrides

Ketones react with compounds having more reactive carbonyl groups such as acid chloride or anhydrides. For example, the reaction of acetone with Ac_2O in presence of BF_3 gives 2,4-pentanedione.

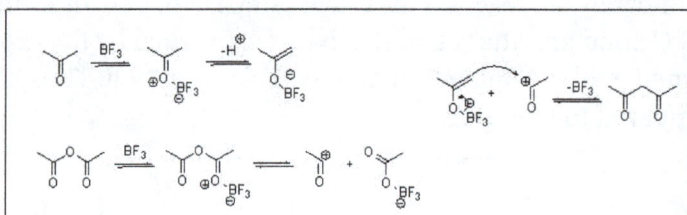

Reactions of 2-Methylpyridine and Analogue

Compounds such as 2- and 4-methylpyridines, and 2- and 4-methylquinolines react with aldehydes in presence of $ZnCl_2$ which catalyzes their conversion into the nitrogen-analogue of an enol. For example, 2-methylpyridine proceeds reaction with benzaldehyde in presence of $ZnCl_2$.

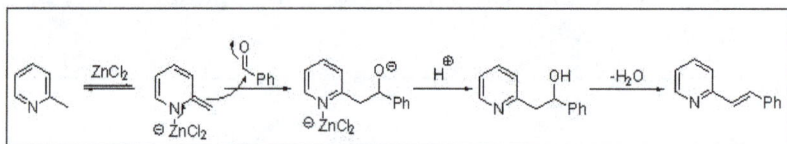

Self-condensation

Self-condensation is an organic reaction in which a chemical compound containing a carbonyl group acts both as the electrophile and the nucleophile in an aldol condensation. It is also called a symmetrical aldol condensation as opposed to a mixed aldol condensation in which the electrophile and nucleophile are different species.

For example, two molecules of acetone condense to a single compound mesityl oxide in the presence of an ion exchange resin:

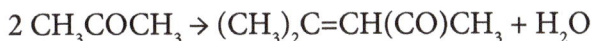

$$2 \ CH_3COCH_3 \rightarrow (CH_3)_2C{=}CH(CO)CH_3 + H_2O$$

For synthetic uses, this is generally an undesirable, but spontaneous and favored side-reaction of mixed aldol condensation, and special precautions are needed to prevent it.

Preventing Self-condensation

In many cases, self-condensation is an unwanted side-reaction. Therefore, chemists have adopted many ways to prevent this from occurring when performing a crossed aldol reaction.

The Use of a More Reactive Electrophile, and a Non-enolizable Partner

If acetophenone and benzaldehyde are put together in the presence of aqueous NaOH, only one product is formed:

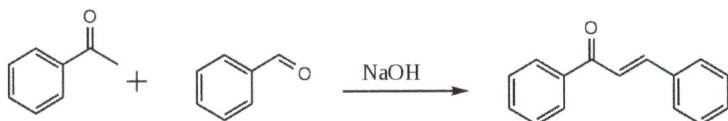

This occurs because benzaldehyde lacks any enolizable protons, so it cannot form an enolate, and the benzaldehyde is much more electrophilic than any unenolized acetophenone in solution. Therefore, the enolate formed from acetophenone will always preferentially attack the benzaldehyde over another molecule of acetophenone.

Making Enolate Ion Quantitatively

When nitromethane and acetophenone are combined using aqueous NaOH, only one product is formed:

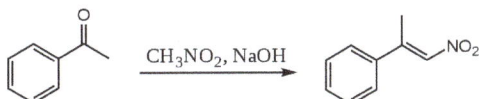

Here, the acetophenone never gets a chance to condense with itself, because the nitromethane is so much more acidic that the nitro "enolate" is made quantitatively.

A similar process can also be used to prevent self-condensation between two ketones. In this case, however, the base used needs to be more powerful. A common base used is Lithium diisopropyl amide (LDA). Here it is used in order to perform the crossed condensation between acetone and cyclohexanone.

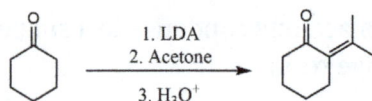

The deprotonation step using LDA is so fast that the enolate formed never gets a chance to react with any unreacted molecules of cyclohexanone. Then the enolate reacts quickly with acetone.

Silyl Enol Ether Formation

Using LDA will not work when attempting to make enolate ion from aldehydes. They are so reactive that self-condensation will occur. One way to get around this is to turn the aldehyde into a silyl enol ether using trimethylsilyl chloride and a base, such as triethylamine, and then perform the aldol condensation. Here this tactic is employed in the condensation of ethanal and benzaldehyde. A Lewis acid, such as $TiCl_4$, must be used in order to promote condensation.

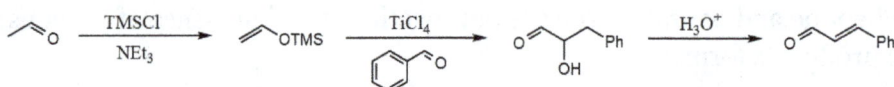

Friedel-Crafts Reactions

Friedel and Crafts reactions are associated with the acylation and alkylation of aromatic systems in presence of Lewis acids.

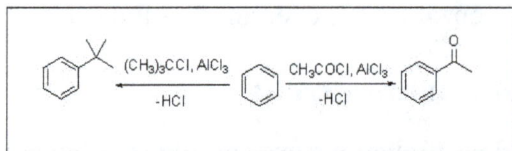

Analogous reactions also occur with alkenes, although they are not widely useful.

Alkylation

The simplest example is the reaction of t -butyl chloride with ethylene to afford neohexyl chloride at -10°C in presence of $AlCl_3$.

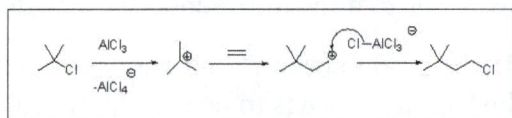

However, several side reactions are usually encountered in alkylation. First, alkene often isomerizes in presence of AlCl$_3$. Second, alkyl halide may rearrange. In addition, the halide produced may undergo further reaction. These aspects limit their applications.

Acylation

The acylation of alkene can be brought using acid chloride or anhydride as acylating agent in presence of Lewis acid. The electrophilic reagent can be either a complex of the acid chloride and Lewis acid or an acylium cation.

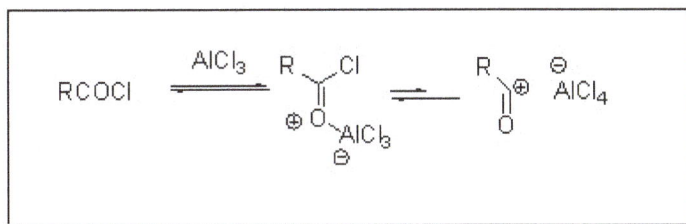

The reaction may end with the uptake of chloride ion to give a b -chloroketone or by elimination to give β, γ-unsaturated ketone which rearranges using the released acid to yield the more stable conjugated ketone.

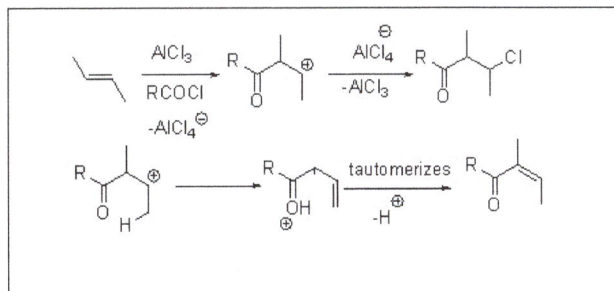

Acylation also suffers due to rearrangement of the alkenes in presence of AlCl$_3$. Thus, the use of less vigorous Lewis acid such as SnCl$_4$ will be useful to minimize the side reactions.

Prins Reaction

The Original Prins Reaction

Prins reaction has remarkable scope which sometimes affords useful molecules that are difficult to prepare by other methods. The reaction involves an electrophilic addition (Markovnikov) of aldehyde or ketone to an alkene in presence of an acid followed by capture of a nucleophile. The outcome of the reaction depends on nature of the substrates and reaction conditions. For examples, with water and protic acid, the reaction product is a 1,3-diol. In absence of water, dehydration takes place to give an allyl-alcohol. With an excess of HCHO, the reaction product is dioxane. When water is replaced by acetic acid, corresponding esters are formed.

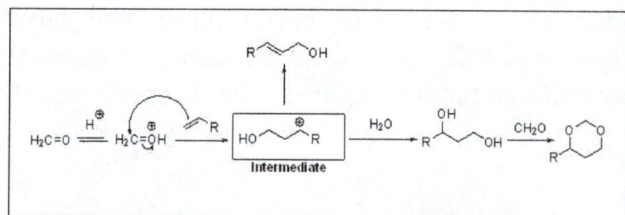

The Formation of Tetrahydropyrans

Aliphatic alkenes undergo Prins reaction. For example, the reaction of propene, HCHO and HCl gives 4-chloro-tetrahydropyran in good yield.

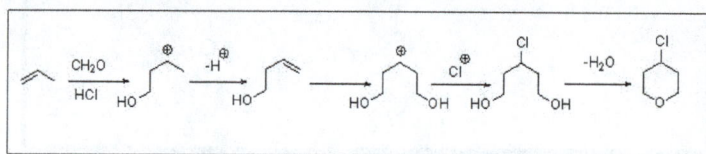

Diastereoselective version of this reaction has also been considerably explored. For example , the synthesis of highly substituted 4-hydroxytetrahydropyran can be accomplished using rhenium(VII) complex, $O_3ReOSiPh_3$, as a catalyst under milder conditions.

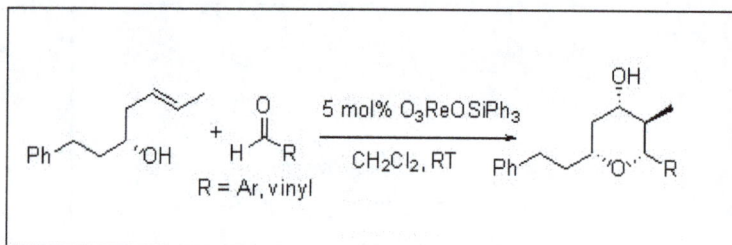

Examples

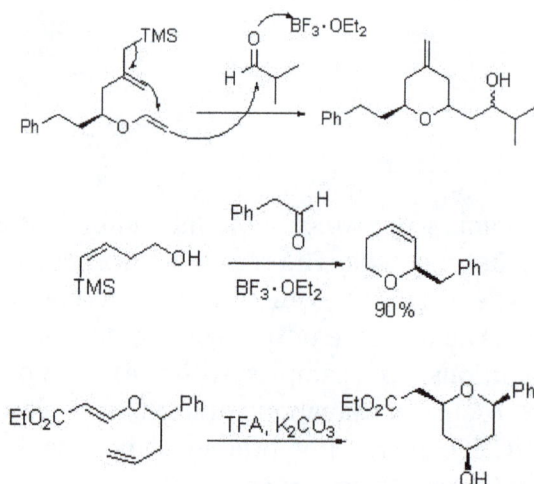

The Carbonyl-Ene Mechanism

The formation of terminal alkene D takes place compared to the more stable internal alkene B. These results suggest that the mechanism is similar to that of a carbonyl ene reaction having the hydrogen transfer and addition of HCHO concerted C. The intermediate C is polar having partial charges E that could be stabilized and the reaction accelerated by protic acids G and Lewis acids I.

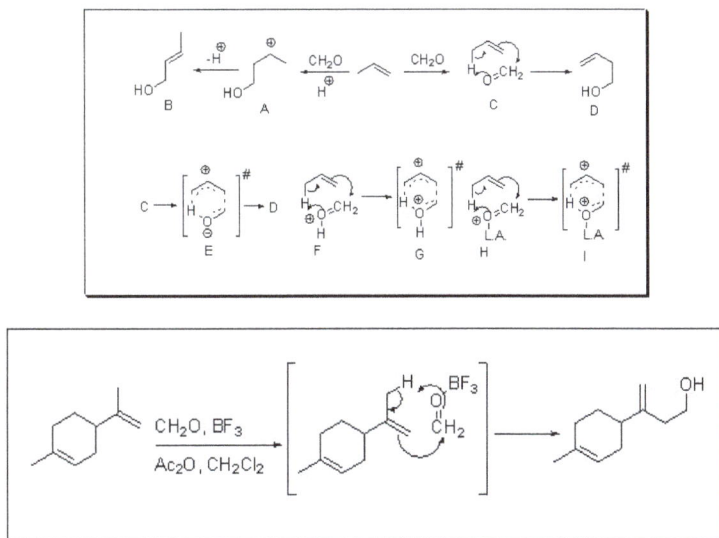

Lewis acids are found to be excellent catalysts and the reaction can be stopped after the first step. This is advancement in the Prins reaction because otherwise a mixture of products is generally obtained. For example, the addition of HCHO to limonene using BF_3 selectively affords a single compound in good yield with excellent chemo- and regioselectivity.

Stereoselectivity

The stereochemistry of the product can be controlled. For example, a mixture of E and Z 1,4-diphenylbut-2-enes with formaldehyde in presence of a mixture of $MeAlCl_2$ and Me_2AlCl gives cyclized product with anti-selectivity in 50% yield. Irrespective of the geometry of the starting alkene a single anti product is obtained. This reaction involves a carbonyl ene reaction followed by Friedel-Crafts alkylation.

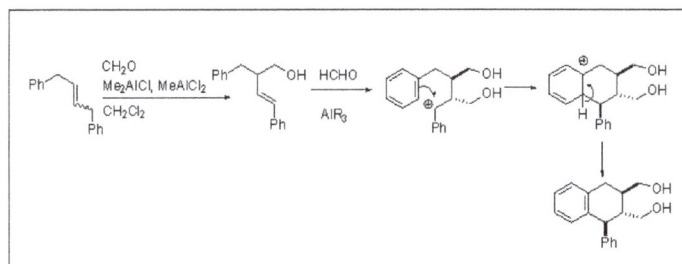

Mannich Reaction

Compounds that are enolic react with a mixture of an aldehydes and a primary or secondary amine in the presence of acid.

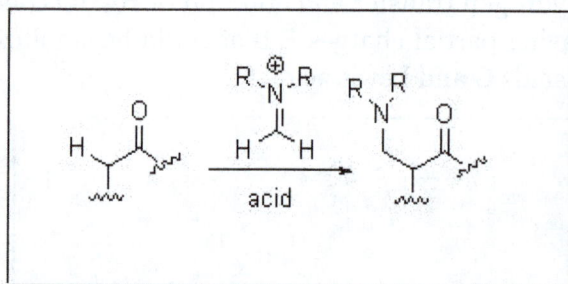

Amines react with aldehydes in the presence of acid to give an adduct that eliminates water to form an electrophiles, and the acid also catalyzes the conversion of ketone into its enol tautomer. The enol then reacts with the electrophiles and the resulting adduct tautomerizes to the amine salt.

Mechanism

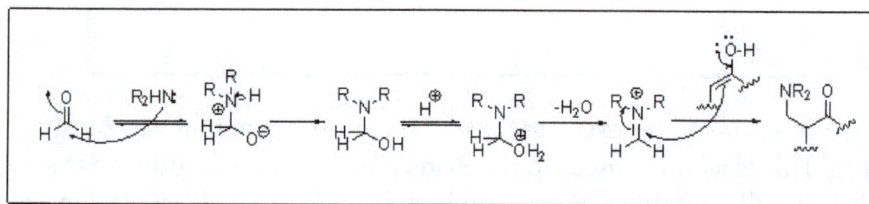

Mannich reaction is one of important reactions in organic synthesis. It provides a simple route for the preparation of amino-ketones and many drug molecules belong to this class. Furthermore, the Mannich products can be converted into enones. For examples, the most reliable method for making enone is to alkylate the Mannich base with MeI and then treat with base.

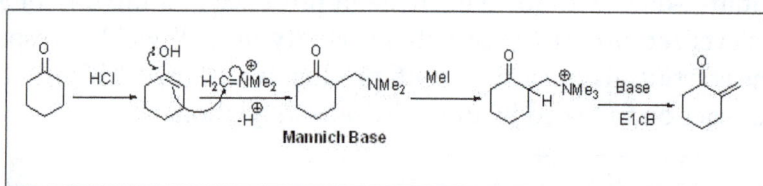

Although the reaction is similar to the Hofmann elimination of a quarternary ammonium salt, Mannich base--hydrochlorides undergo elimination readily because the double-bond which is generated is conjugated with unsaturated group. Eliminations of this type have advantages. For example, the reduction of C=C bond gives the next highest homologue of the ketone from which the Mannich base was derived.

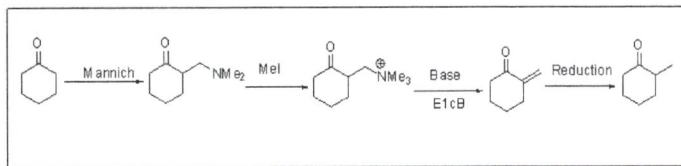

Furthermore, the quarternary salts of Mannich bases are latent sources of α,β -unsaturated carbonyl compounds required for condensation reactions. For example, for a base-catalyzed reaction utilizing but-3-en-2-one, it is better to employ the quarternized Mannich base derived from acetone, from which the unsaturated ketone is generated in situ by the action of a base, than to use the free but-3-en-2-one which is prone to undergo polymerization. The best example of this application is the Robinson annelation in steroid synthesis.

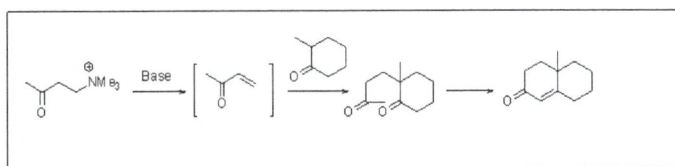

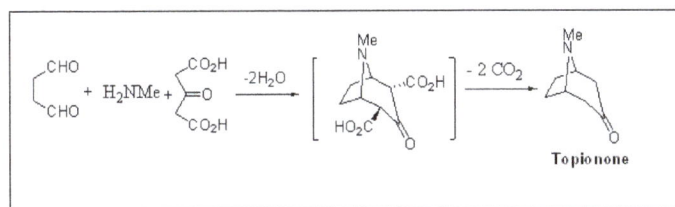

Mannich reaction finds huge applications in alkaloid synthesis. For example, the Robinson's synthesis of tropionone from calcium acetonedicarboxylate, methylamine and succindialdehyde gives the product in 90% yield. It consists of two Mannich reactions, followed by spontaneous decarboxylation of the dibasic β -keto acid.

Examples

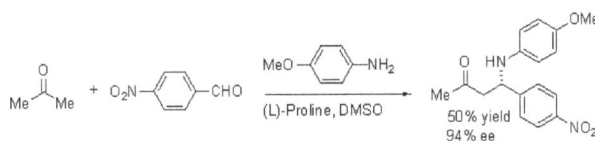

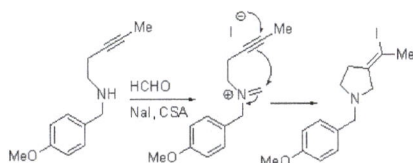

The Friedel–Crafts reactions are a set of reactions developed by Charles Friedel and James Crafts in 1877 to attach substituents to an aromatic ring. There are two main types of Friedel–Crafts reactions: alkylation reactions and acylation reactions. Both proceed by electrophilic aromatic substitution. The general reaction figure is shown below.

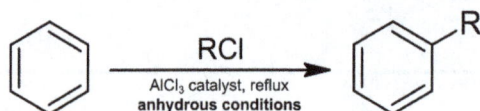

Friedel–Crafts alkylation

Template:Reaction-box Friedel–Crafts alkylation involves the alkylation of an aromatic ring with an alkyl halide using a strong Lewis acid catalyst. With anhydrous ferric chloride as a catalyst, the alkyl group attaches at the former site of the chloride ion. The general mechanism is shown below.

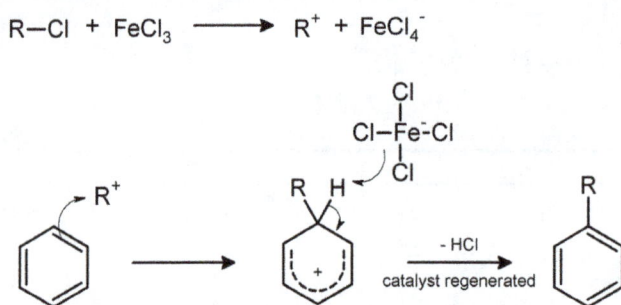

This reaction has one big disadvantage, namely that the product is more nucleophilic than the reactant due to the electron donating alkyl-chain. Therefore, another hydrogen atom is substituted with an alkyl-chain, which leads to overalkylation of the molecule. Also, if the chloride ion is not attached to a tertiary carbon atom or secondary carbon atom, then the carbocation formed (R^+) will undergo a carbocation rearrangement reaction. This reactivity is due to the relative stability of the tertiary and secondary carbocation over the primary carbocations.

Steric hindrance can be exploited to limit the number of alkylations, as in the *t*-butylation of 1,4-dimethoxybenzene.

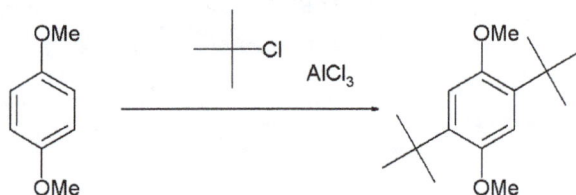

Alkylations are not limited to alkyl halides: Friedel–Crafts reactions are possible with any carbocationic intermediate such as those derived from alkenes and a protic acid, Lewis acid, enones, and epoxides. An example is the synthesis of neophyl chloride from benzene and methallyl chloride:

$$H_2C=C(CH_3)CH_2Cl + C_6H_6 \rightarrow C_6H_5C(CH_3)_2CH_2Cl$$

In one study the electrophile is a bromonium ion derived from an alkene and NBS:

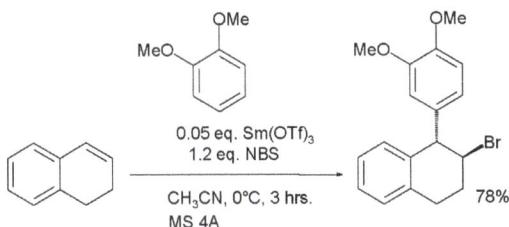

In this reaction samarium(III) triflate is believed to activate the NBS halogen donor in halonium ion formation.

Friedel–Crafts Dealkylation

Friedel–Crafts alkylation is a reversible reaction. In a reversed Friedel–Crafts reaction or Friedel–Crafts dealkylation, alkyl groups can be removed in the presence of protons and a Lewis acid.

For example, in a multiple addition of ethyl bromide to benzene, *ortho* and *para* substitution is expected after the first monosubstitution step because an alkyl group is an activating group. However, the actual reaction product is 1,3,5-triethylbenzene with all alkyl groups as a *meta* substituent. Thermodynamic reaction control makes sure that thermodynamically favored *meta* substitution with steric hindrance minimized takes prevalence over less favorable *ortho* and *para* substitution by chemical equilibration. The ultimate reaction product is thus the result of a series of alkylations and dealkylations.

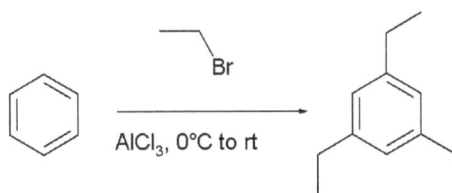

Friedel–Crafts Acylation

Friedel–Crafts acylation is the acylation of aromatic rings with an acyl chloride using a strong Lewis acid catalyst. Friedel–Crafts acylation is also possible with acid anhydrides. Reaction conditions are similar to the Friedel–Crafts alkylation mentioned above. This reaction has several advantages over the alkylation reaction. Due to the electron-withdrawing effect of the carbonyl group, the ketone product is always less reactive than the original molecule, so multiple acylations do not occur. Also, there are no carbocation rearrangements, as the carbonium ion is stabilized by a resonance structure in which the positive charge is on the oxygen.

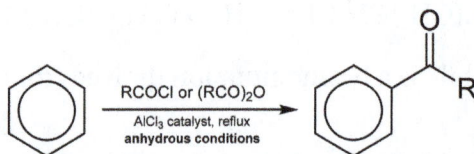

The viability of the Friedel–Crafts acylation depends on the stability of the acyl chloride reagent. Formyl chloride, for example, is too unstable to be isolated. Thus, synthesis of benzaldehyde via the Friedel–Crafts pathway requires that formyl chloride be synthesized *in situ*. This is accomplished via the Gattermann-Koch reaction, accomplished by treating benzene with carbon monoxide and hydrogen chloride under high pressure, catalyzed by a mixture of aluminium chloride and cuprous chloride.

Reaction Mechanism

In a simple mechanistic view, the first step consists of dissociation of a chloride ion to form an acyl cation (acylium ion):

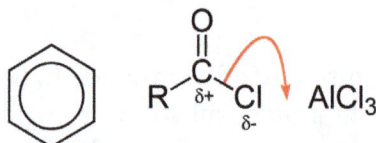

In some cases, the Lewis acid binds to the oxygen of the acyl chloride to form an adduct. Regardless, the resulting acylium ion or a related adduct is subject to nucleophilic attack by the arene:

Finally, chloride anion (or $AlCl_4^-$) deprotonates the ring (an arenium ion) to form HCl, and the $AlCl_3$ catalyst is regenerated:

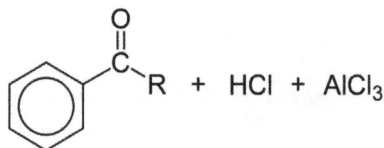

If desired, the resulting ketone can be subsequently reduced to the corresponding alkane substituent by either Wolff–Kishner reduction or Clemmensen reduction. The net result is the same as the Friedel–Crafts alkylation except that rearrangement is not possible.

Friedel–Crafts Hydroxyalkylation

Arenes react with certain aldehydes and ketones to form the hydroxyalkylated product for example in the reaction of the mesityl derivative of glyoxal with benzene to form a benzoin with an alcohol rather than a carbonyl group:

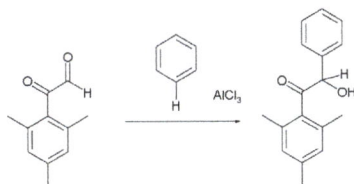

Friedel–Crafts Sulfonylation

Under Friedel–Crafts reaction conditions, arenes react with sulfonyl halides and sulfonic acid anhydrides affording sulfones. Commonly used catalysts include $AlCl_3$, $FeCl_3$, $GaCl_3$, BF_3, $SbCl_5$, $BiCl_3$ and $Bi(OTf)_3$, among others. Intramolecular Friedel–Crafts cyclization occurs with 2-phenyl-1-ethanesulfonyl chloride, 3-phenyl-1-propanesulfonyl chloride and 4-phenyl-1-butanesulfonyl chloride on heating in nitrobenzene with $AlCl_3$. Sulfenyl and sulfinyl chlorides also undergo Friedel–Crafts–type reactions, affording sulfides and sulfoxides, respectively. Both aryl sulfinyl chlorides and diaryl sulfoxides can be prepared from arenes through reaction with thionyl chloride in the presence of catalysts such as $BiCl_3$, $Bi(OTf)_3$, $LiClO_4$ or $NaClO_4$.

Scope and Variations

This reaction is related to several classic named reactions:

- The acylated reaction product can be converted into the alkylated product via a Clemmensen reduction.

- The Gattermann–Koch reaction can be used to synthesize benzaldehyde from benzene.

- The Gatterman reaction describes arene reactions with hydrocyanic acid.

- The Houben–Hoesch reaction describes arene reactions with nitriles.

- A reaction modification with an aromatic phenyl ester as a reactant is called the Fries rearrangement.

- In the Scholl reaction two arenes couple directly (sometimes called Friedel–Crafts arylation).

- In the Zincke–Suhl reaction p-cresol is alkylated to a cyclohexadienone with tetrachloromethane.

- In the Blanc chloromethylation a chloromethyl group is added to an arene with formaldehyde, hydrochloric acid and zinc chloride.

- The Bogert–Cook Synthesis (1933) involves the dehydration and isomerization of *1-β-phenylethylcyclohexanol* to the octahydro derivative of phenanthrene

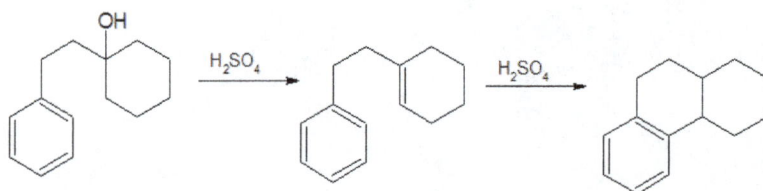

- The Darzens–Nenitzescu Synthesis of Ketones (1910, 1936) involves the acylation of cyclohexene with acetyl chloride to methylcyclohexenylketone.

- In the related Nenitzescu reductive acylation (1936) a saturated hydrocarbon is added making it a reductive acylation to methylcyclohexylketone

- Nencki Reaction (1881) is the ring acetylation of phenols with acids in the presence of zinc chloride.

- In a green chemistry variation aluminium chloride is replaced by graphite in an alkylation of p-xylene with 2-bromobutane. This variation will not work with primary halides from which less carbocation involvement is inferred.

Dyes

Friedel–Crafts reactions have been used in the synthesis of several triarylmethane and xanthene dyes. Examples are the synthesis of thymolphthalein (a pH indicator) from two equivalents of thymol and phthalic anhydride:

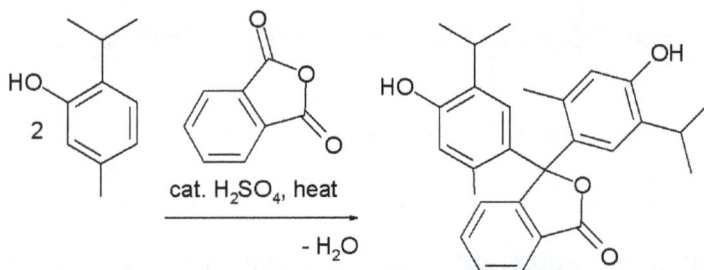

A reaction of phthalic anhydride with resorcinol in the presence of zinc chloride gives the fluorophore Fluorescein. Replacing resorcinol by N,N-diethylaminophenol in this reaction gives rhodamine B:

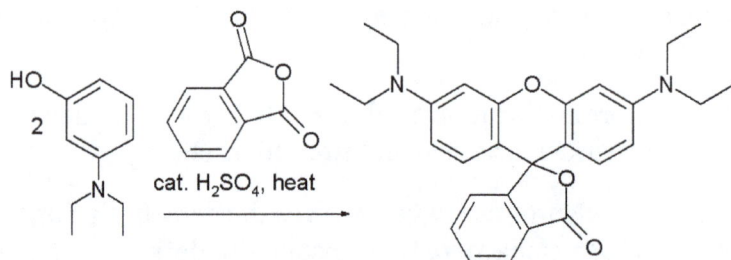

Haworth Reactions

The Haworth reaction is a classic method for the synthesis of 1-tetralone. In it benzene is reacted with succinic anhydride, the intermediate product is reduced and a second FC acylation takes place with addition of acid.

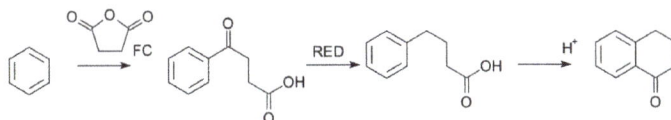

In a related reaction, phenanthrene is synthesized from naphthalene and succinic anhydride in a series of steps.

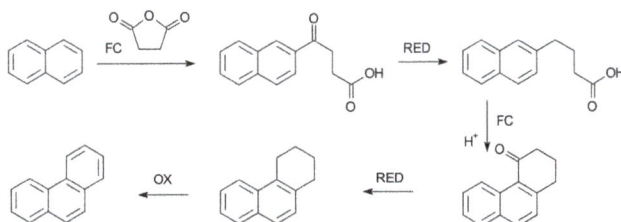

Friedel–Crafts Test for Aromatic Hydrocarbons

Reaction of chloroform with aromatic compounds using an aluminium chloride catalyst gives triarylmethanes, which are often brightly colored, as is the case in triarylmethane dyes. This is a bench test for aromatic compounds.

References

- Goren, M. B.; Sokoloski, E. A.; Fales, H. M. (2005). "2-Methyl-(1Z,3E)-butadiene-1,3,4-tricarboxylic Acid, "Isoprenetricarboxylic Acid"". Journal of Organic Chemistry. 70 (18): 7429–7431. PMID 16122270. doi:10.1021/jo0507892

- Carey, Francis A.; Sundberg, Richard J. (1993). Advanced Organic Chemistry Part B Reactions and Synthesis (3rd ed.). 233 Spring Street, NY: Plenum. p. 55. ISBN 0-306-43440-7

- Claisen, L.; Claparède, A. (1881). "Condensationen von Ketonen mit Aldehyden". Berichte der Deutschen Chemischen Gesellschaft. 14 (1): 2460–2468. doi:10.1002/cber.188101402192

- Varela, J. A.; Gonzalez-Rodriguez, C.; Rubin, S. G.; Castedo, L.; Saa, C. (2006). "Ru-Catalyzed Cyclization of Terminal Alkynals to Cycloalkenes". Journal of the American Chemical Society. 128 (30): 9576–9577. PMID 16866480. doi:10.1021/ja0610434

- Smith, M. B.; March, J. (2001). Advanced Organic Chemistry (5th ed.). New York: Wiley Interscience. pp. 1218–1223. ISBN 0-471-58589-0

- Mukaiyama T. (1982). "The Directed Aldol Reaction". Organic Reactions. 28: 203–331. doi:10.1002/0471264180.or028.03

- Lockner, James. "Stoichiometric Enamine Chemistry" (PDF). Baran Group, The Scripps Research Institute. Retrieved 26 November 2014

- Lambert, T. H.; Danishefsky, S. J. (2006). "Total Synthesis of UCS1025A". Journal of the American Chemical Society. 128 (2): 426–427. doi:10.1021/ja0574567

- Carey, Francis A.; Sundberg, Richard J. (1993). Advanced Organic Chemistry Part A: Structure and Mechanisms (3rd ed.). 233 Spring Street, New York, N.Y.: Plenum. p. 458. ISBN 0-306-43440-7

- Reformatsky, S. (1887). "Neue Synthese zweiatomiger einbasischer Säuren aus den Ketonen". Berichte der Deutschen Chemischen Gesellschaft. 20 (1): 1210–1211. doi:10.1002/cber.188702001268

- Wiener, Jake. "Enantioselective Organic Catalysis:Non-MacMillan Approaches" (PDF). Re-trieved 29 November 2014

- Tung, C. C.; Speziale, A. J.; Frazier, H. W. (June 1963). "The Darzens Condensation. II. Reaction of Chloroacetamides with Aromatic Aldehydes". The Journal of Organic Chemistry. 28 (6): 1514–1521. doi:10.1021/jo01041a018

- March, Jerry (1985), Advanced Organic Chemistry: Reactions, Mechanisms, and Structure (3rd ed.), New York: Wiley, ISBN 0-471-85472-7

- Weiss, U.; Edwards, J. M. (1968). "A one-step synthesis of ketonic compounds of the pentalane, [3,3,3]- and [4,3,3]-propellane series". Tetrahedron Letters. 9 (47): 4885. doi:10.1016/S0040-4039(00)72784-5

- Lockner, James. "Stoichiometric Enamine Chemistry" (PDF). Baran Group, The Scripps Research Institute. Retrieved 26 November 2014

- Perkin, W. H. (1868). "On the artificial production of coumarin and formation of its homologues". Journal of the Chemical Society. 21: 53–61. doi:10.1039/js8682100053

- Enamines: Synthesis: Structure, and Reactions, Second Edition, Gilbert Cook (Editor). 1988, Marcel Dekker, NY. ISBN 0-8247-7764-6

- Ballester, Manuel (April 1955). "Mechanisms of The Darzens and Related Condensations Manuel Ballester". Chemical Reviews. 55 (2): 283–300. doi:10.1021/cr50002a002

- Perkin, W. H. (1877). "On some hydrocarbons obtained from the homologues of cinnamic acid; and on anethol and its homologues". Journal of the Chemical Society. 32: 660–674. doi:10.1039/js8773200660

- March Jerry; (1985). Advanced Organic Chemistry reactions, mechanisms and structure (3rd ed.). New York: John Wiley & Sons, inc. ISBN 0-471-85472-7

- Dippy, J. F. J.; Evans, R. M. (1950). "The nature of the catalyst in the Perkin condensation". J. Org. Chem. 15 (3): 451–456. doi:10.1021/jo01149a001

- Capon, Brian; Wu, Zhen Ping (April 1990). "Comparison of the tautomerization and hydrolysis of some secondary and tertiary enamines". The Journal of Organic Chemistry. 55 (8): 2317–2324. doi:10.1021/jo00295a017

- Claisen, L. (1887). "Ueber die Einführung von Säureradicalen in Ketone". Berichte der Deutschen Chemischen Gesellschaft. 20 (1): 655–657. doi:10.1002/cber.188702001150

Substitution Reaction of Carbon-Nitrogen Bonds

This section focuses on the formation of carbon-nitrogen bonds. This bond can be categorized into the reaction caused by nucleophilic nitrogen with electrophilic carbon and the reaction of electrophilic nitrogen with nucleophilic carbon. The aspects elucidated in this chapter are of vital importance, and provide a better understanding of carbon-nitrogen bonds formation.

Carbon-Nitrogen Bonds Formation

Principles

The methods for the formation of bonds between nitrogen and aliphatic carbon can be broadly divided into two categories: (i) reaction of nucleophilic nitrogen with electrophilic carbon, and (ii) reaction of electrophilic nitrogen with nucleophilic carbon. In this section, we will try to cover some of the important reactions.

Ritter Reaction

The Ritter reaction is a chemical reaction that transforms a nitrile into an *N*-alkyl amide using various electrophilic alkylating reagents. The original reaction formed the alkylating agent using an alkene in the presence of a strong acid: The reaction has been the subject of several literature reviews.

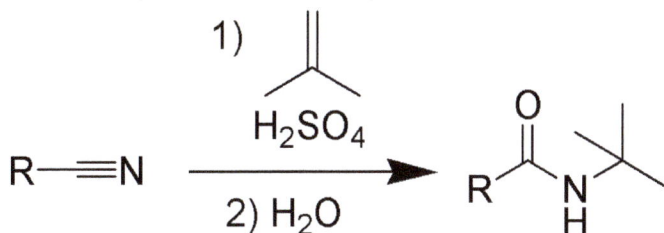

Primary, secondary, tertiary, and benzylic alcohols, as well as *tert*-butyl acetate, also successfully react with nitriles in the presence of strong acids to form amides via the Ritter reaction.

History

The Ritter reaction is named after John J. Ritter, an American chemist who received his Ph.D. from Columbia University. In 1948, P. Paul Minieri, Ritter's student, submitted work on the reaction as his Ph.D. thesis. The reaction still has significance today due to its applicability and reproducibility of amides via stabilized carbocations.

Reaction Mechanism

The Ritter reaction proceeds by the electrophilic addition of either the carbenium ion 2 or covalent species to the nitrile. The resulting nitrilium ion 3 is hydrolyzed by water to the desired amide 5.

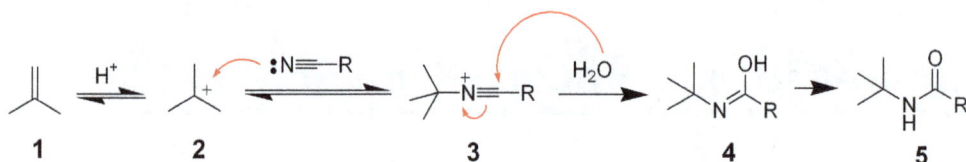

Applications

The Ritter reaction is most useful in the formation of amides in which the nitrogen has a tertiary alkyl group. It is also used in industrial processes as it can be effectively scaled up from laboratory experiments to large-scale applications while maintaining high yield. Real world applications include Merck's industrial-scale synthesis of anti-HIV drug Crixivan (indinavir); the production of the falcipain-2 inhibitor PK 11195; the synthesis of the alkaloid aristotelone; and synthesis of Amantadine, an antiviral and antiparkinsonian drug. Other applications of the Ritter reaction include synthesis of dopamine receptor ligands and production of amphetamine from allylbenzene.

A problem with the Ritter reaction is the necessity of an extremely strong acid catalyst in order to produce the carbocation. This corrosive type of chemical poses an environmental hazard for chemical waste and safety risk for running the reaction itself. However, other methods have been proposed in order promote carbocation formation, including photosensitized electron transfer or direct photolysis.

Treatment of a tertiary alcohol or the corresponding alkene with concentrated sulphuric acid and a nitrile affords an amide that in acidic conditions undergoes hydrolysis to give amine. First, a

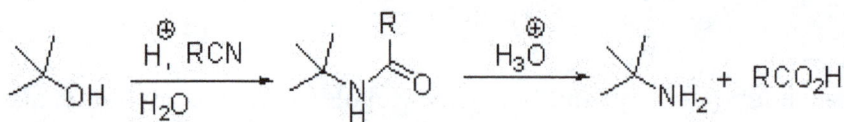

Mechanism

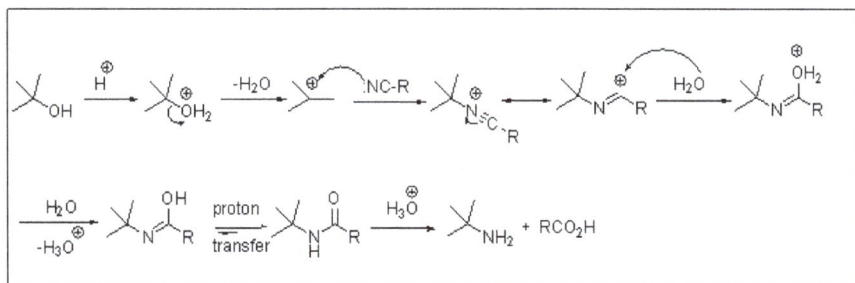

Tertiary carbocation is formed which is attacked by the nitrile to give a quaternary ion. The latter is decomposed by water to provide an amide which, in acidic conditions, is hydrolyzed to give an amine.

Examples

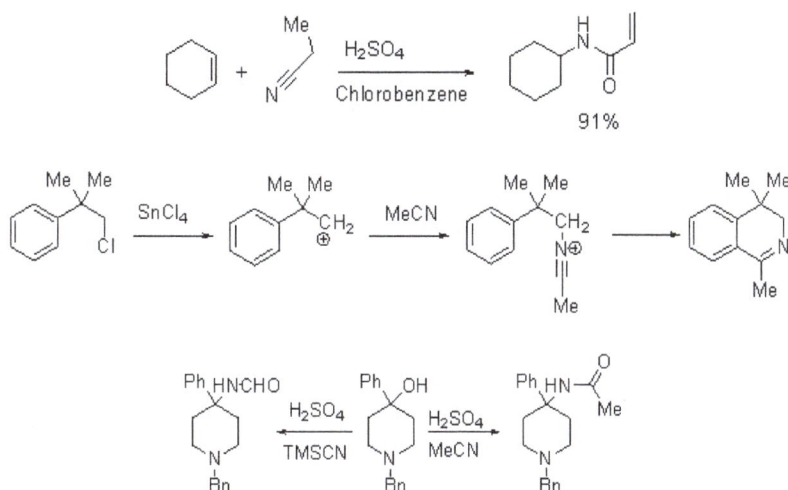

Gabriel Synthesis

The Gabriel synthesis is a chemical reaction that transforms primary alkyl halides into primary amines. Traditionally, the reaction uses potassium phthalimide. The reaction is named after the German chemist Siegmund Gabriel.

The Gabriel reaction has been generalized to include the alkylation of sulfonamides and imides, followed by deprotection, to obtain amines.

The alkylation of ammonia is often an unselective and inefficient route to amines. In the Gabriel method, phthalimide anion is employed as a surrogate of H_2N^-.

Traditional Gabriel Synthesis

In this method, the sodium or potassium salt of phthalimide is *N*-alkylated with a primary alkyl halide to give the corresponding *N*-alkylphthalimide. The reaction fails with most secondary alkyl halides:

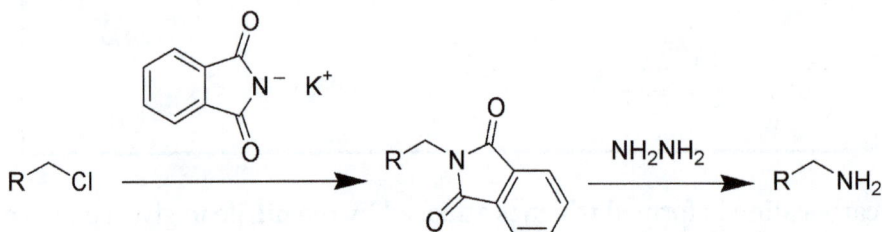

Upon workup by acidic hydrolysis the primary amine is liberated as the amine salt. Alternatively the workup may be via the Ing–Manske procedure, involving reaction with aqueous or ethanolic hydrazine at reflux. This method produces a precipitate of phthalhydrazide along with the primary amine. The first technique often produces bad yields or side products; separation of phthalhydrazide can be unpleasant. For these reasons, other methods for liberating the amine from the phthalimide exist. Even with the use of the hydrazinolysis method, the Gabriel method suffers from relatively harsh conditions.

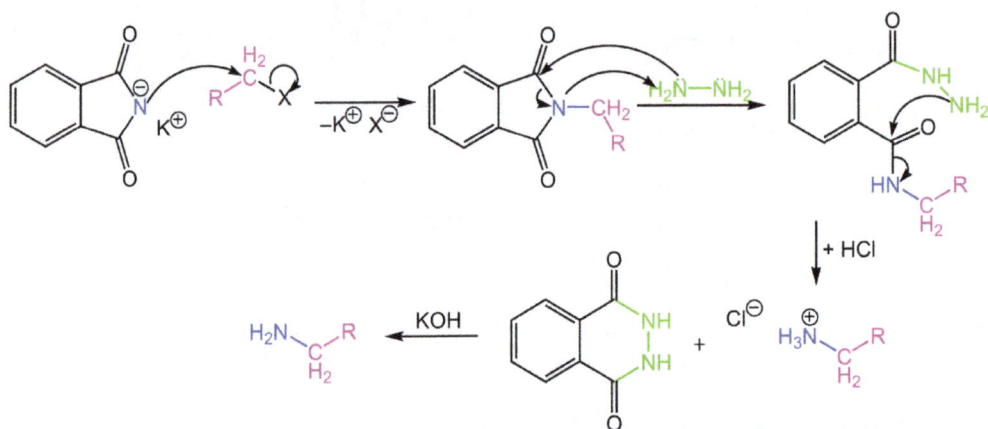

Mechanism of the Gabriel synthesis

Alternative Gabriel Reagents

Many alternative reagents have been developed to complement the use of phthalimides. Most such reagents, e.g. the sodium salt of saccharin, and di-tert-butyl-iminodicarboxylate. These reagents are electronically similar to the phthalimide salts, consisting of imido nucleophiles.. In terms of their advantages, these reagents hydrolyze more readily, extend the reactivity to secondary alkyl halides, and allow the production of secondary amines.

Gabriel method provides an effective route for the synthesis primary amine. In this reaction, phthalimide, having an acidic N-H group, reacts with base to afford a nitrogen
containing anion that, as a nucleophile, undergoes substitution on alkyl halides. The
resulting compound on hydrolysis with alkali gives the primary amine.

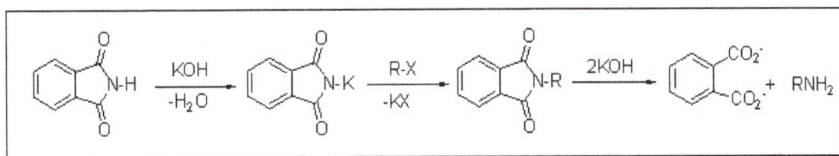

Mechanism

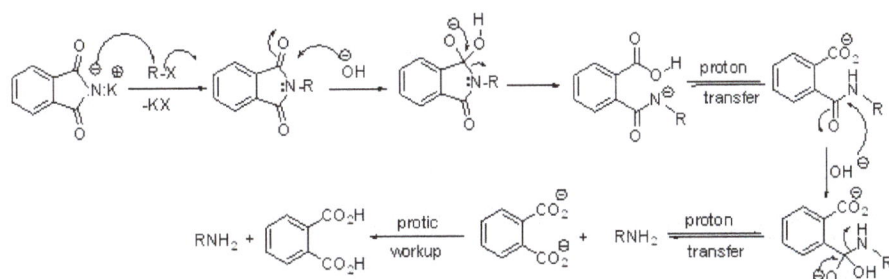

The use of hydrazine to release the primary amine has been subsequently accomplished.
This procedure is called Manske modification which finds more useful because it is gen-
tle to other functional groups.

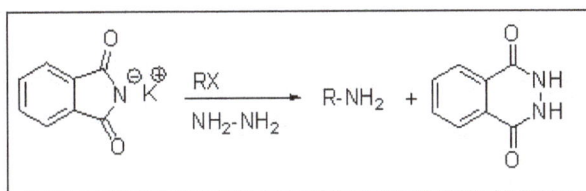

Examples

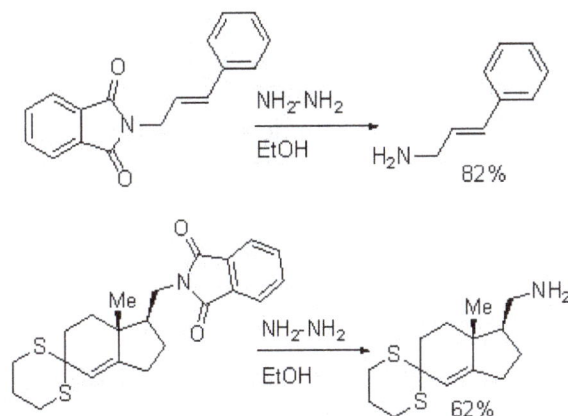

Gabriel–Colman Rearrangement

The Gabriel–Colman rearrangement is the chemical reaction of a saccharin or phthalimido ester with a strong base, such as an alkoxide, to form substituted isoquinolines. This rearrangement, a ring expansion, is seen to be general if there is an enolizable hydrogen on the group attached to the nitrogen, since it is necessary for the nitrogen to abstract a hydrogen to form the carbanion that will close the ring. As shown in the case of the general example below, X is either CO or SO_2.

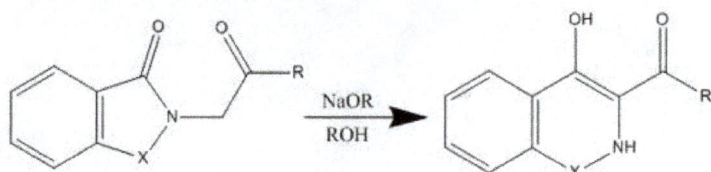

Mechanism

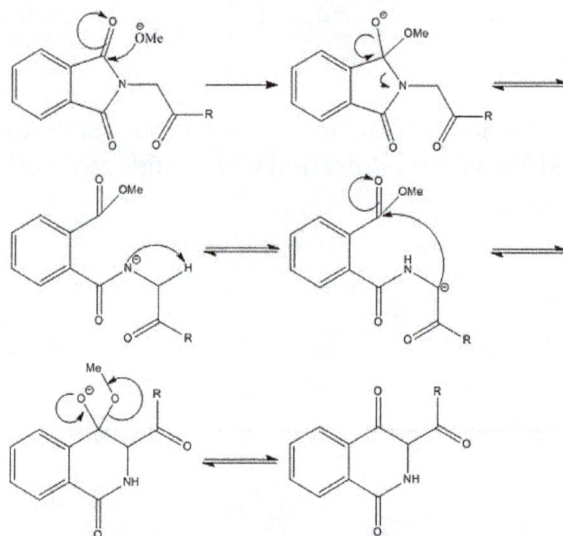

The reaction mechanism starts with an attack on the carbonyl group by a strong base, such as methoxide ion. The ring is then opened, forming an imide anion. This is then followed by a rapid isomerization of the imide anion to the carbanion. This is facilitated by the electron withdrawing effect of the substituent, which allows for greater stabilization of the adjacent carbanion with respect to the imide anion. The reaction is then completed when the methoxide is displaced by the ring closing, which results in a ring expansion. The rate determining step of this reaction is the attack of the carbanion on the carbomethoxy group.

The displacement of the methoxide is analogous to the displacement seen in the Dieckman condensation, as it is also a result of a ring closure.

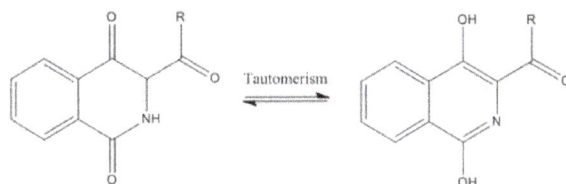

Furthermore, tautomerization can occur on both of the carbonyl groups on the ring, with interconversion of the keto form to the enol form and the amide form to the imidic acid form.

Applications

The major application of the Gabriel–Colman rearrangement is in the formation of iso-quinolines, due to the relatively high yield of the desired products. Therefore, studies in which either the product or an intermediate is an isoquinoline, the Gabriel–Colman rearrangement can be utilized. This reaction has been utilized in the production of in-termediates for the synthesis of potential anti-inflammatory agents. It has also been used in the study of phthalimide and saccharin derivatives as mechanism based inhib-itors for three enzymes; the human leukocyte elastase, cathepsin G and proteinase 3. Phthalimide derivatives were seen to be inactive, while saccharin derivatives were seen to be fair inhibitors of these enzymes.

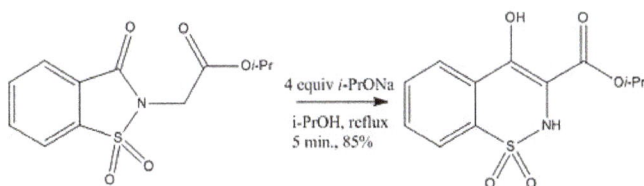

In a study of the derivatives of 3-Oxo-1,2-benzoisothiazoline-2-acetic acid 1,1-dioxide, the Gabriel–Colman rearrangement was employed in the conversion of Isopropyl (1,1-di-oxido-3-oxo-1,2-benzothiazol-2(3H)-yl)acetate to Isopropyl 4-hydroxy-2H-1,2-benzo-thiazine-3-carboxylate 1,1-dioxide, as shown above. This reaction has shown a percent yield of 85%.

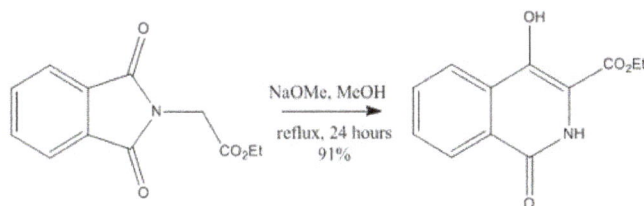

In another study, N-phthalimidoglycine ethyl ester was used to synthesize 4-hydroxy-isoquinoline through use of a Gabriel–Colman rearrangement, as shown above. This reaction has shown a percent yield of 91%. The formation of this product was an im-portant step in the study of the synthesis of 4,4'-functionalized 1,1'-biisoquinolines.

Potassium phthalimide proceeds nucleophilic substitution with α -halo acetate and the resulting product in the presence of base undergoes rearrangement to afford isoquinoline derivatives.

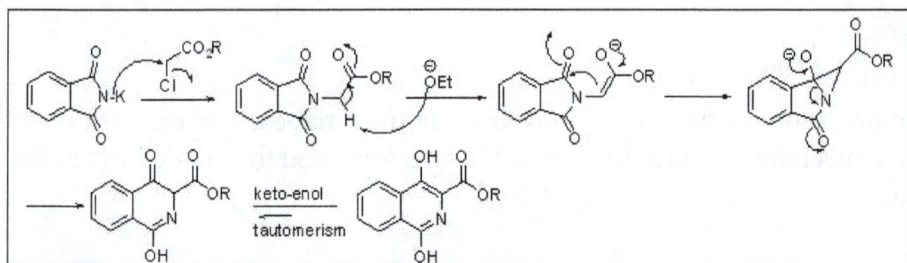

Reactions of Other Nitrogen Nucleophiles

Nitrite

Metal nitrites can react with alkyl halides at both nitrogen and oxygen to give nitro compounds and nitrites, respectively, whose proportions depend on the structure of the reactants and reaction conditions. For example, silver nitrite suspended in ether reacts with alkyl halides to give a mixture of nitro-compounds and nitrites whose proportions depend on the nature of alkyl halides.

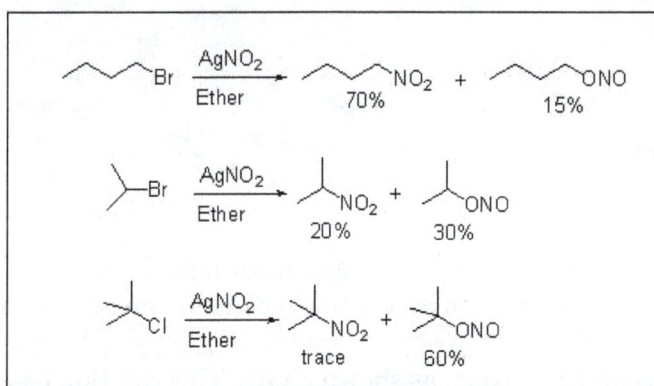

Azide

Azides react with halides to give alkyl azides that could be reduced to afford primary amines.

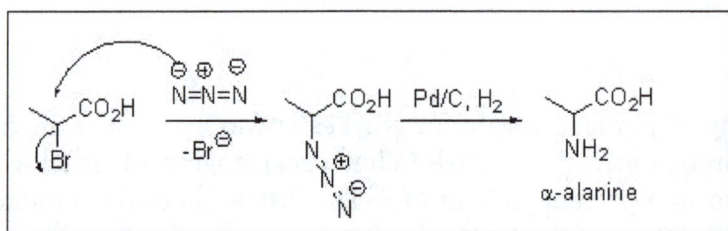

Hydrazine

The reaction of hydrazine with alkyl halides generally gives dialkylated products. This is because the introduction of the first alkyl group increases the nucleophilicity of the alkylated nitrogen, so that further alkylation tends to takes place.

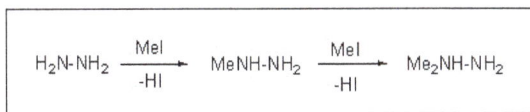

Hofmann bromianton of alkylurea can give manoalkylated hydrazines in good yield.

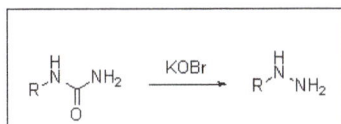

Addition of Nitrogen Nucleophiles to Unsaturated Carbon

Reactions with Aldehydes and Ketones

The condensation of aldehydes with amines finds wide applications. The fate of the adduct depends on the structure of the aldehydes, amine, and the reaction conditions. For examples, formaldehyde reacts with ammonia to give urotropine (hexamethylenetetramine).

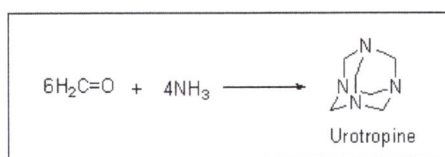

Aromatic aldehydes generally provide condensation products. This strategy has been used to construct stereoregular chiral main chain polymers from optically active di-amines and dialdehydes with excellent yield which are otherwise difficult to access by other methods.

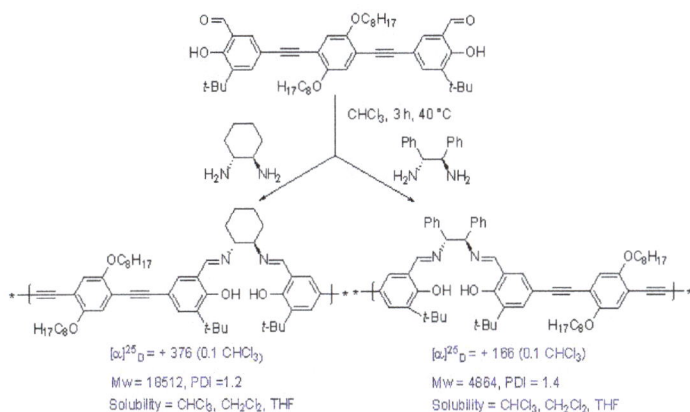

Ugi Reaction

The four-component condensation of isocyanide, a carboxylic acid, an aldehydes or ketone and ammonia or amine gives bisamide. The product formation probably takes place from a reaction between carboxylic acid, the isocyanide, and the imine formed from the aldehydes or ketone and ammonia or the primary amine. The use of N-protected amino acids allows the reaction to be used for peptide synthesis.

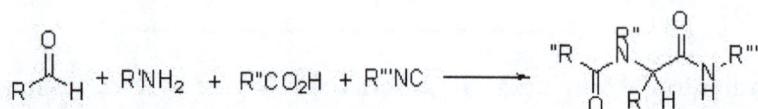

Mechanism

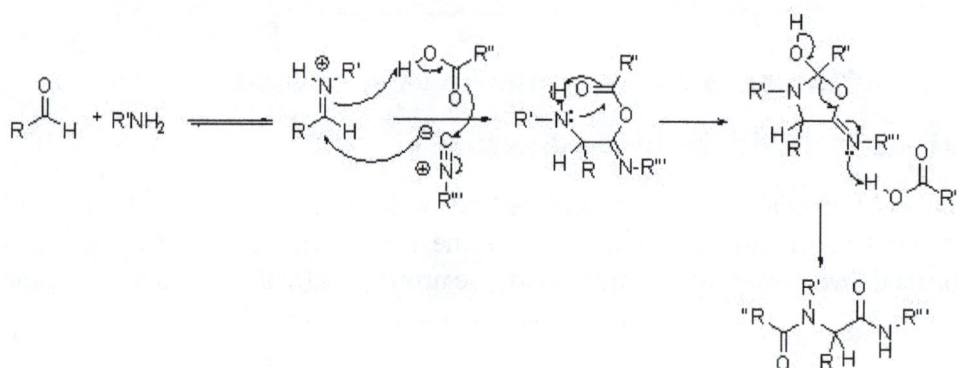

Examples

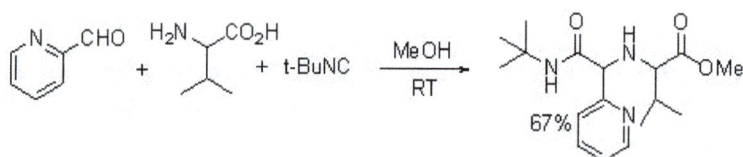

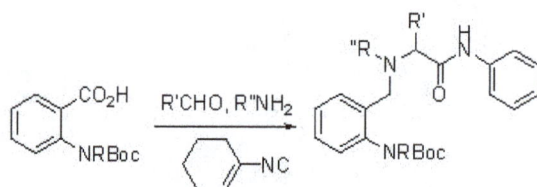

Eschweiler-Clarke (Clark) Methylation

Secondary amines could be readily methylated using the combination of formaldehyde and formic acid under heating. This process has been extensively used in total synthesis.

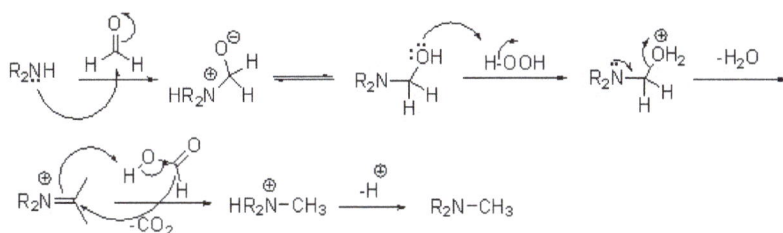

Examples

66%

HCHO, DCO$_2$D   R = CH$_2$D

DCHO, DCO$_2$D   R = CHD$_2$

DCDO, DCO$_2$D   R = CD$_3$

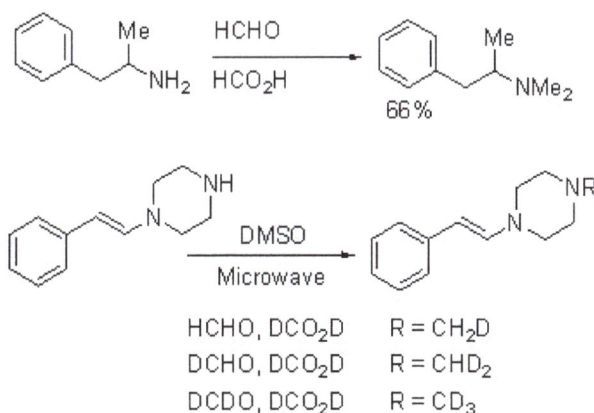

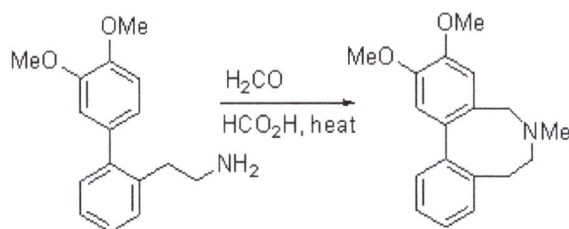

Eschweiler–Clarke Reaction

The Eschweiler–Clarke reaction (also called the Eschweiler–Clarke methylation) is a chemical reaction whereby a primary (or secondary) amine is methylated using excess formic acid and formaldehyde. Reductive amination reactions such as this one will not produce quaternary ammonium salts, but instead will stop at the tertiary amine stage. It is named for the German chemist Wilhelm Eschweiler (1860–1936) and the British chemist Hans Thacher Clarke (1887-1972).

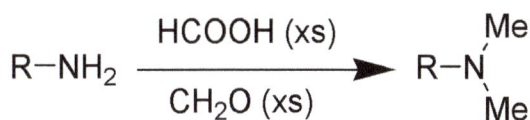

Mechanism

The first methylation of the amine begins with imine formation with formaldehyde. The formic acid acts as a source of hydride and reduces the imine to a secondary amine. The driving force is the formation of the gas carbon dioxide. Formation of the tertiary amine is similar, but slower due to the difficulties in iminium ion formation.

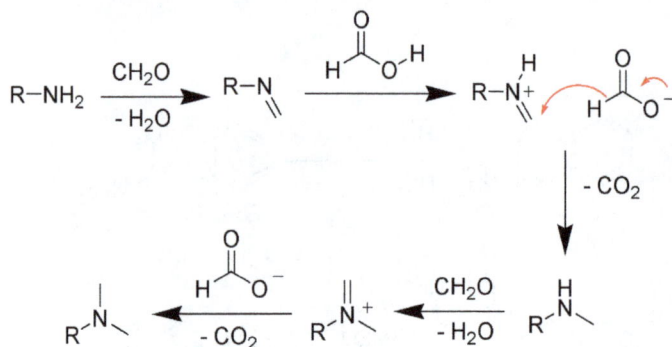

From this mechanism it is clear that a quaternary ammonium salt will never form, because it is impossible for a tertiary amine to form another imine or iminium ion.

Chiral amines typically do not racemize under these conditions.

Altered versions of this reaction replace formic acid with sodium cyanoborohydride.

Robinson-Schopf Reaction

Compounds that are enolic or potentially enolic react with a mixture of aldehydes and primary or secondary amine in the presence of an acid to afford amine salt which, after basification, gives an aminomethyl derivative. The reaction has found wide applications in organic synthesis. For example, the synthesis of tropinone can be accomplished in 90% yield. This reaction follows biomimetic approach to forming alkaloids.

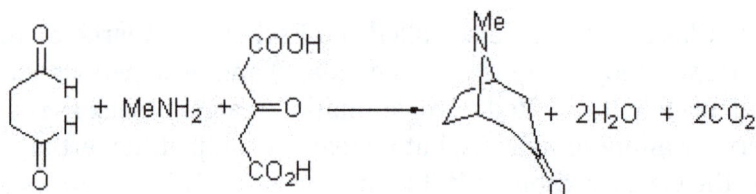

Mechanism

The synthesis involves two Mannich reactions followed by spontaneous decarboxylation of the dibasic β-keto acid.

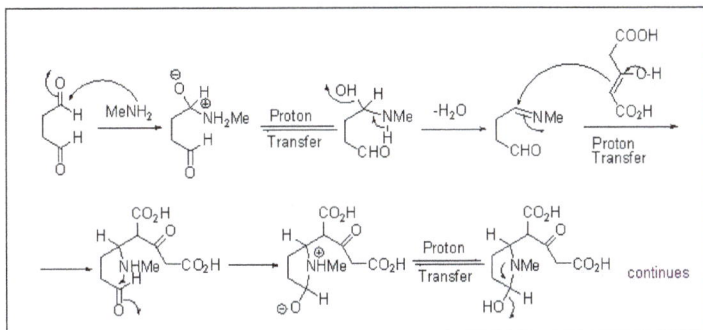

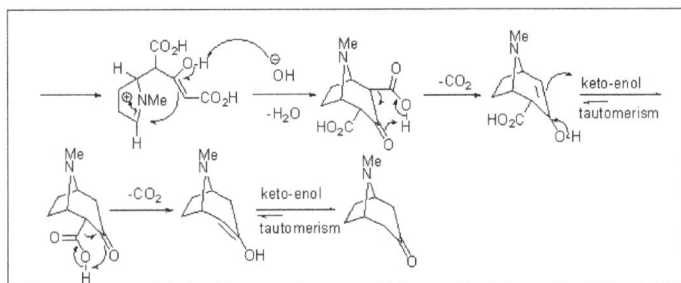

Examples

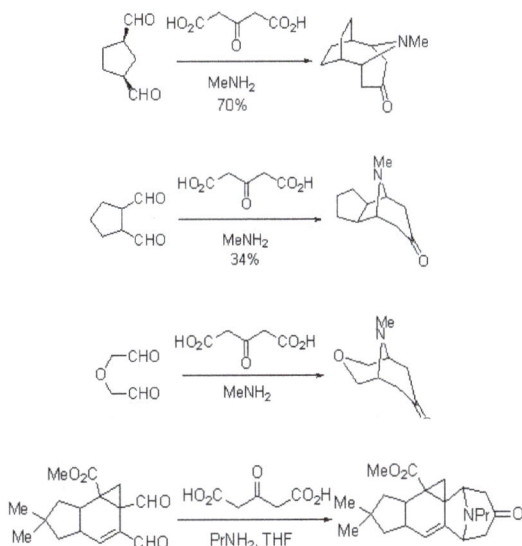

The Strecker Synthesis

The condensation of an aldehyde with amine gives imine that can undergo reaction with cyanide ion in situ to give an α -aminonitrile which on hydrolysis gives α -amino acid. This process constitutes a useful method for α -amino acid synthesis. Asymmetric version has also been explored.

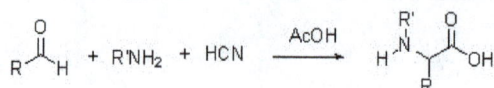

Mechanism

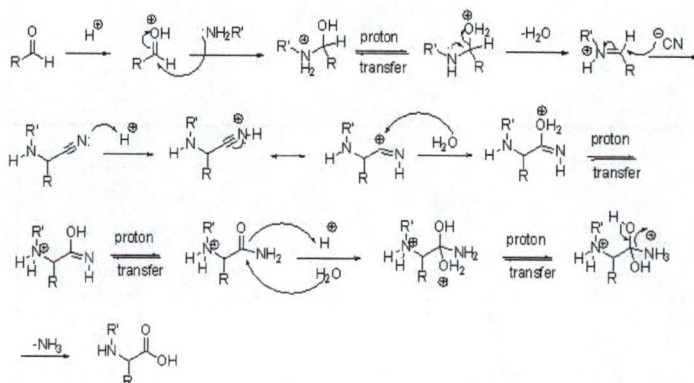

Examples

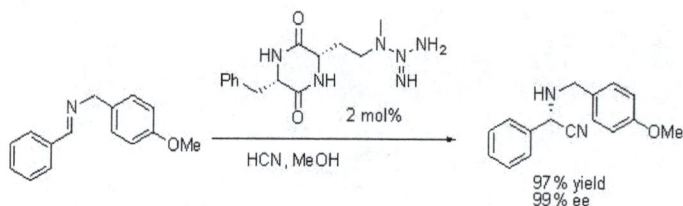

97 % yield
99 % ee

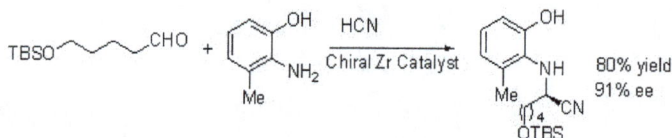

80% yield
91% ee

Stork Enamine Synthesis

Enamines are specific enol equivalents to alkylate aldehydes and ketones. They are formed when aldehydes or ketones react with secondary amines.

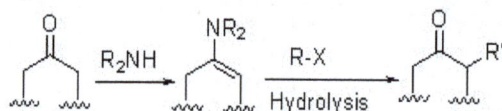

Mechanism

The mechanism shows how they react with alkylating agent to form a new carbon-carbon bond.

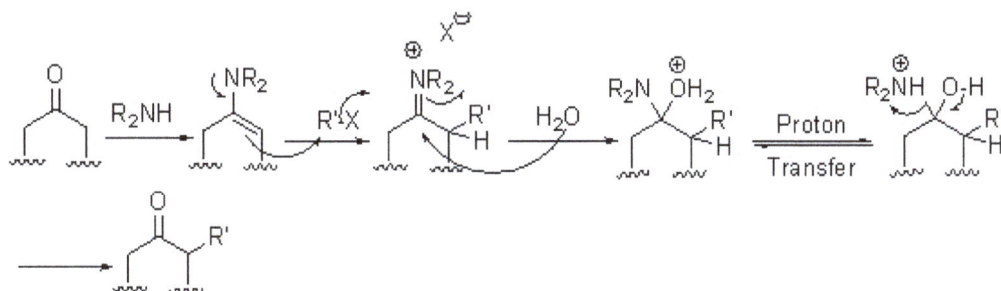

The overall process amounts to an enolate alkylation. The reaction conditions are mild and no self-condensation is observed.

Examples

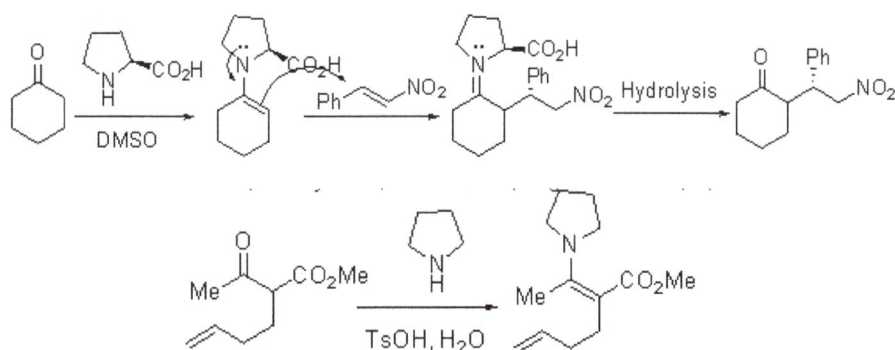

Gabriel-Cromwell Reaction

Amines react with α -bromoacrylates to give aziridines in the presence of base.

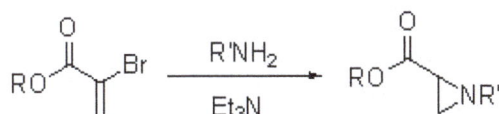

Mechanism

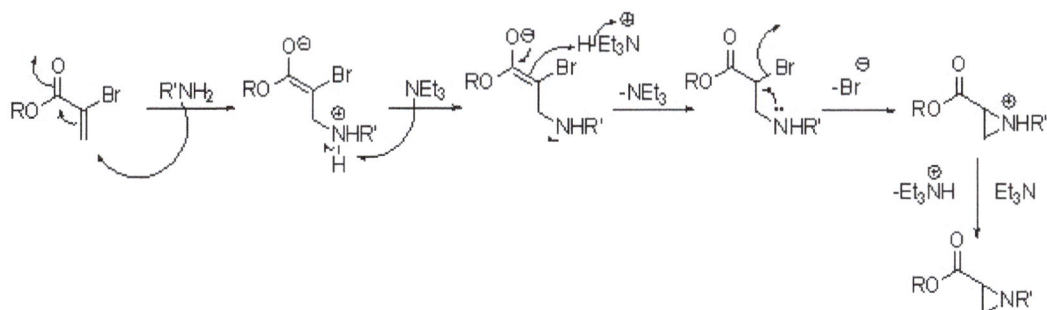

Examples

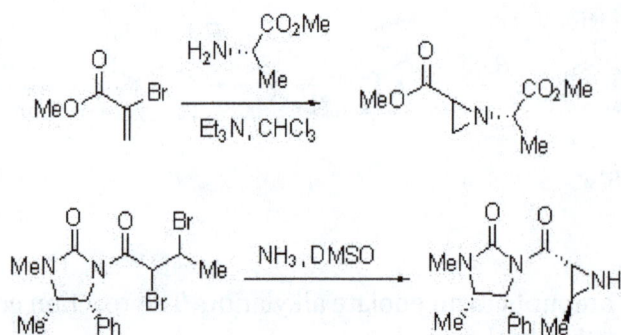

Schweizer Allyl Amine Synthesis

This reaction involves the combined use of Gabriel and Wittig chemistry for the synthesis of allyl amines from phthalimide, vinyl phosphonium salt and aldehyde in the presence of base.

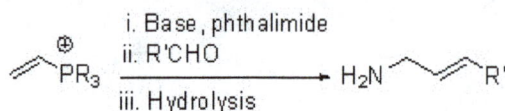

Mechanism

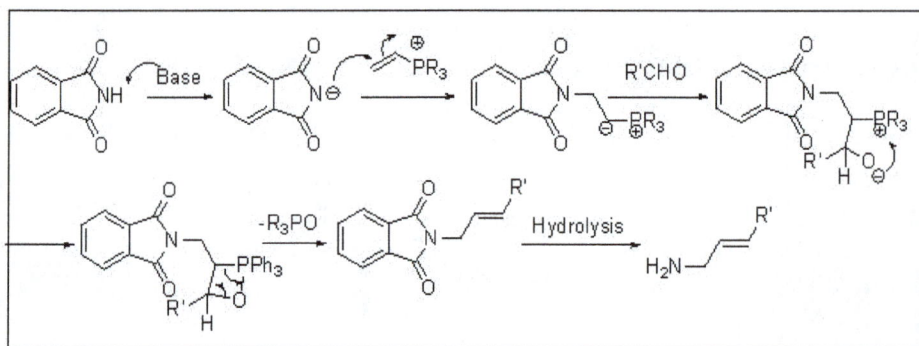

Borche Reduction

Aldehydes and ketones react with amines to give imine that could be reduced using MCNBH$_3$(M= Li, Na) to amines.

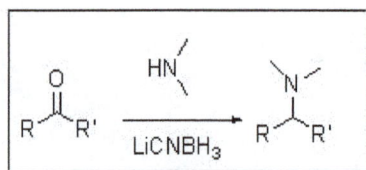

Mechanism

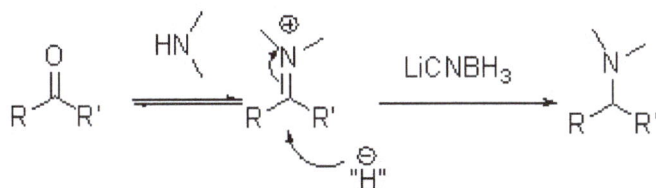

The success of this method rests on the much greater reactivity of imine salt compared to carbonyl group of aldehydes and ketone to the reducing agent.

Examples

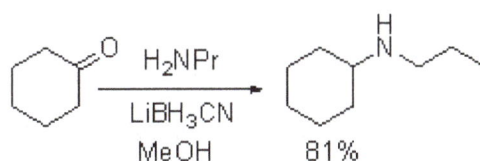

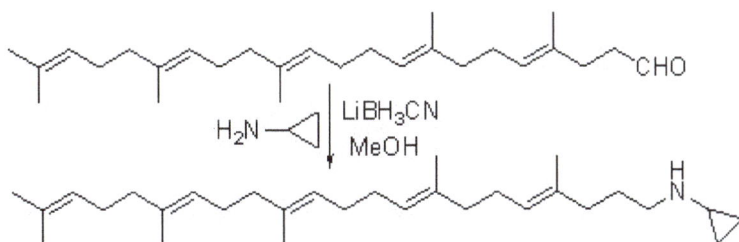

Doebner Reaction (Beyer Synthesis)

Aryl amine reacts with aldehydes and enolizable carbonyl compounds via condensation followed by aromatic electrophilic substitution and autoxidation to give quinolines.

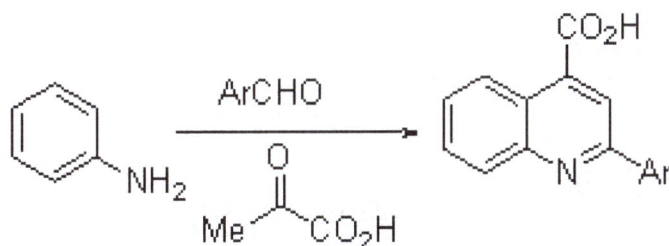

Mechanism

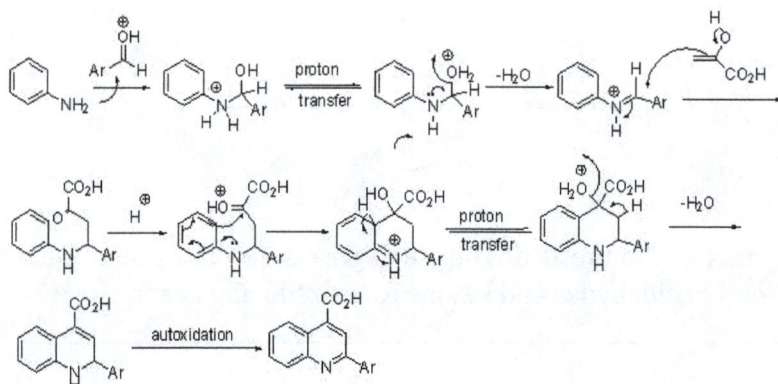

Examples

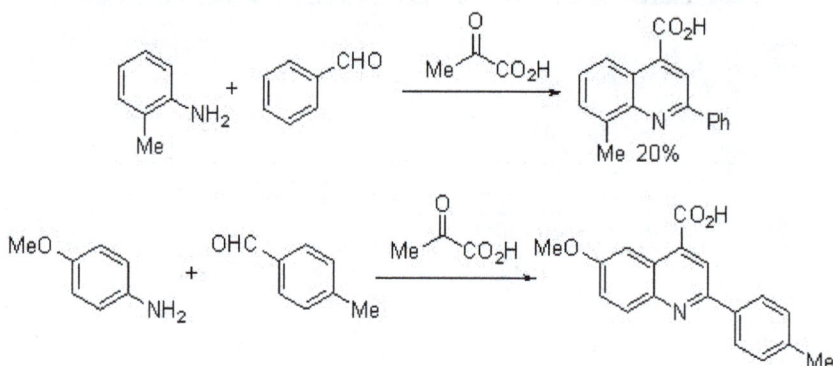

Strecker Amino acid Synthesis

The Strecker amino acid synthesis, also known simply as the Strecker synthesis, was devised by German chemist Adolph Strecker, and is a term used for a series of chemical reactions that synthesize an amino acid from an aldehyde or ketone. The aldehyde is condensed with ammonium chloride in the presence of potassium cyanide to form an α-aminonitrile, which is subsequently hydrolyzed to give the desired amino acid. In the original Strecker reaction acetaldehyde, ammonia, and hydrogen cyanide combined to form after hydrolysis alanine.

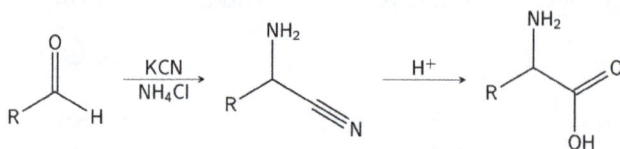

While usage of ammonium salts gives unsubstituted amino acids, primary and secondary amines also successfully give substituted amino acids. Likewise, the usage of ketones, instead of aldehydes, gives α,α-disubstituted amino acids.

The traditional synthesis of Adolph Strecker from 1850 gives racemic α-amino nitriles, but several procedures utilizing asymmetric auxiliaries or asymmetric catalysts have been developed.

Reaction Mechanism

In the first part of the reaction, the carbonyl oxygen of an aldehyde is protonated, followed by a nucleophilic attack of ammonia to the carbonyl carbon. After subsequent proton exchange, water is cleaved from the iminium ion intermediate. A cyanide ion then attacks the iminium carbon yielding an aminonitrile.

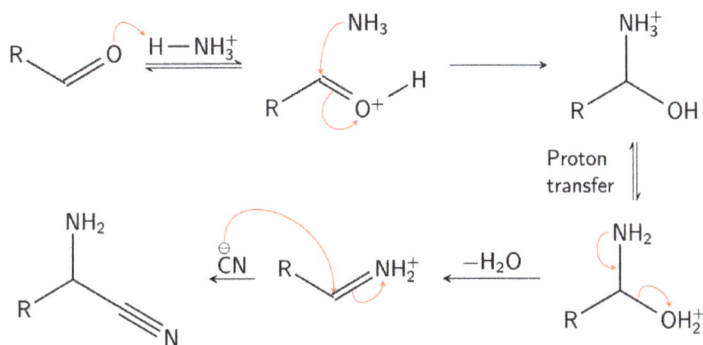

In the second part of the Strecker Synthesis the nitrile nitrogen of the aminonitrile is protonated, and the nitrile carbon is attacked by a water molecule. A 1,2-diamino-diol is then formed after proton exchange and a nucleophilic attack of water to the former nitrile carbon. Ammonia is subsequently eliminated after the protonation of the amino group, and finally the deprotonation of a hydroxyl group produces an amino acid.

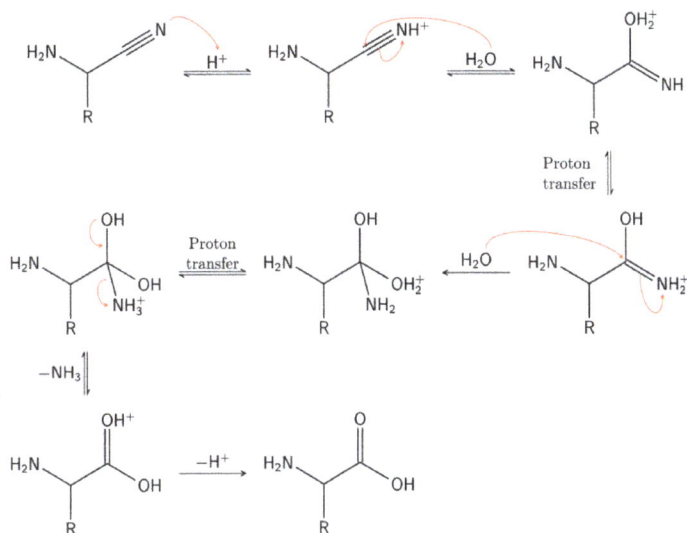

One example of the Strecker synthesis is a multikilogram scale synthesis of an L-valine derivative starting from Methyl isopropyl ketone:

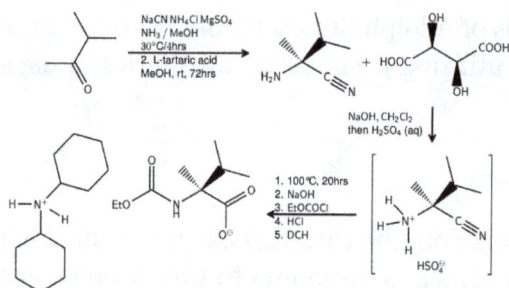

Asymmetric Strecker Reactions

The asymmetric Strecker reaction was pioneered by Kaoru Harada in 1963. By replacing ammonia with (S)-alpha-phenylethylamine as chiral auxiliary the ultimate reaction product was chiral alanine. The first asymmetric synthesis via a chiral catalyst was reported in 1996.

Catalytic Asymmetric Strecker Reactions

Catalytic asymmetric Strecker reaction can be effected using thiourea-derived catalyst. In 2012, a BINOL-derived catalyst was employed to generate chiral cyanide anion.

Substitution by Nuclophilic Nitrogen at Unsaturated Carbon

These reactions are analogous to the base-catalyzed hydrolysis of the carbonyl compounds. The order of reactivity of the compounds having different leaving groups is: acid halide > anhydride > ester > amide.

Reactions of Ammonia and Amines

- Amines readily react with acylating agents such as acid chlorides and acid anhydrides in the presence of base to give amides. These are the general practical methods used for the acylation reactions.

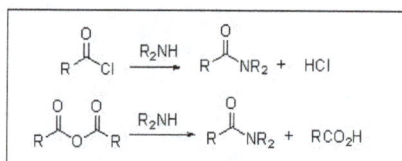

- The condensation of carboxylic acid with amines using DCC is also a popular route for the amide formation which will be discussed in the section on peptide synthesis. DCC is converted into urea which can be separated by filtration.

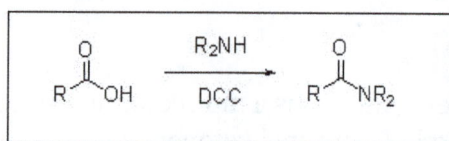

- Amides are weak nucleophiles and the further acylation occurs very slowly that leads to isolation of the monoacylated product. However, intramolecular displacement can take place if a five or six membered ring formation is possible. For example, heating of the acyclic diamide of succinic acid affords succinimide in good yield.

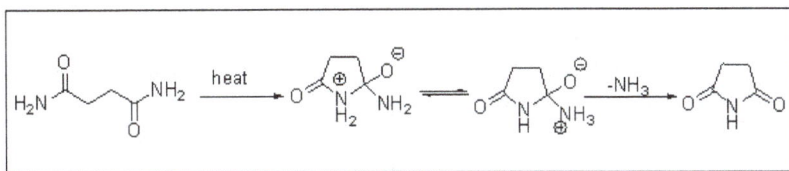

- Cyclic imides can be easily prepared (practical method) by heating a mixture of acid anhydride and amine.

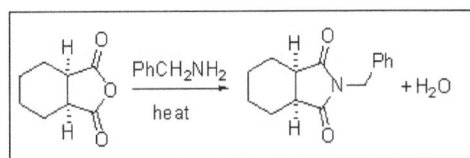

- The reaction of amines with ethyl chloroformate and carbonyl chloride affords urethanes and ureas, respectively.

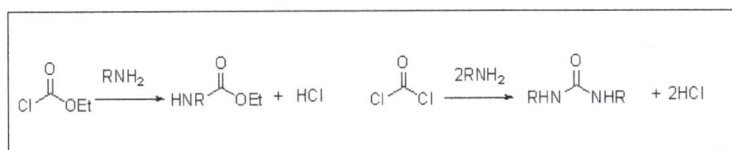

Reactions of Other Nitrogen Nucleophiles

- Acid chloride, anhydrides and esters react with hydrazine to provide acid hydrazides which are less nucleophilic and the further acylation requires a more vigorous conditions. Thus, monoacylated product can be obtained in high yield.

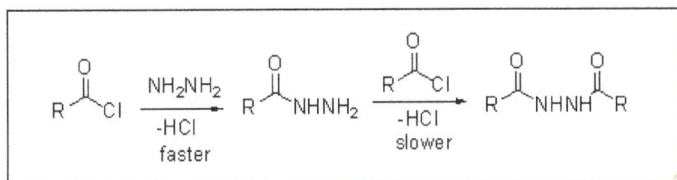

- Hydroxylamine reacts with carboxylic acid derivatives giving hydroxyamic acids.

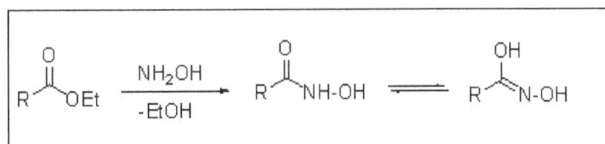

- Reaction of acid chloride with sodium azide gives acid azide which is the substrate precursor for Curtius rearrangement.

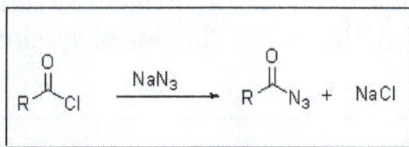

Reactions of Electrophilic Nitrogen

Four electrophilic nitrogen containing groups, nitroso (-NO), nitro (-NO$_2$), arylazo (-N=NAr), and arylimino (=NAr), could be bonded to aliphatic carbon via C-N bond formation.

Nitrosation

The reactions of nitroso group may be done by two ways. In first, enols react with nitrous acid or an organic nitrite by an acid-catalyzed process. The electrophile is to be the nitrosonium ion, NO$^+$.

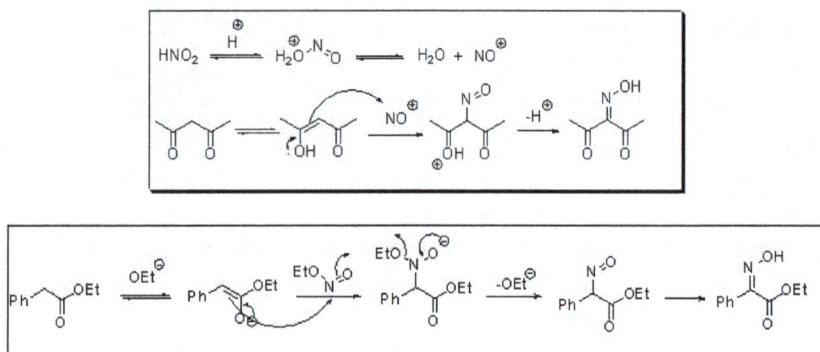

The products have two main synthetic applications. First, several heterocyclic syntheses are performed by reducing β-keto oximes in the presence of compounds with which the products can proceed reaction to give cyclized products such as pyrroles.

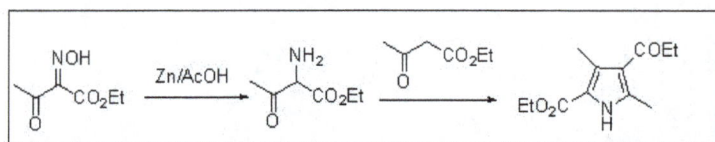

Second, the oxime is hydrolyzed into a carbonyl group.

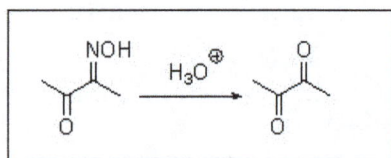

Nitration

Compounds generate enolate react with alkyl nitrate by base-catalyzed procedure. The reaction pathway is similar to that of the base-catalyzed nitrosation process.

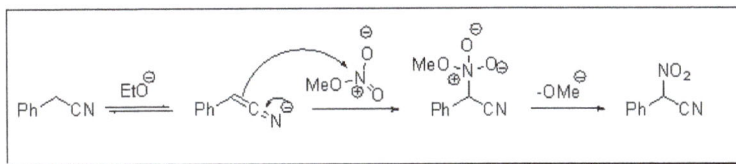

Imine Formation

Compounds that can generate enolates react with aromatic nitroso compounds to give imine which on hydrolysis affords carbonyl group. For example, 2,4-nitrotoluene can be converted into 2,4-dinitrobenzaldehyde by this process.

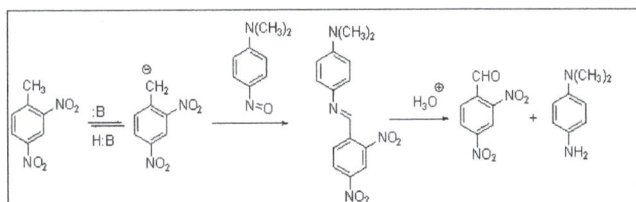

α-Amino Acids, Peptides and Proteins

Peptides and proteins are naturally occurring polymers in living systems. They are derived from α -amino acids linked together by amide bonds. The distinction between peptides and proteins is that the polymers with molecular weights less than 10000 are termed peptides and those with higher molecular weights are termed proteins.

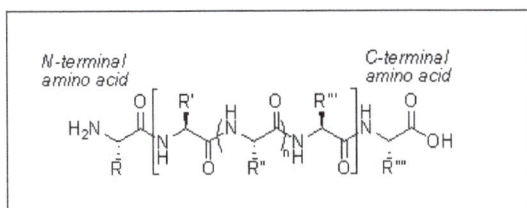

The acid-catalyzed hydrolysis of peptides and proteins provides the constituent α -amino acids which are, except glycine, chiral having L-configuration. The syntheses of peptides followed to date have been based on the reverse of this process. Therefore α -amino acids are of significant importance.

The Synthesis of α -Amino Acids

The following are the some of the common methods used for the synthesis of α -amino acids.

From α-Halo Acids

The simplest method consists of the conversion of carboxylic acid into it's α-bromo-derivative which could be reacted with ammonia to give α-amino acid.

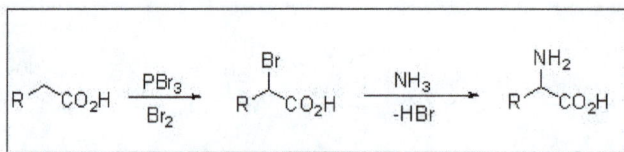

Hell-Volhard-Zelinski reaction is generally employed for the preparation of α-bromo acid. It involves the treatment of the acid with bromine in the presence of a small amount of phosphorus to give acid bromide which undergoes (electrophilic) bromination at the α-position via its enol tautomer. The resulting product exchanges with more of the acid to afford α-bromo acid together with more acid bromide for the further bromination.

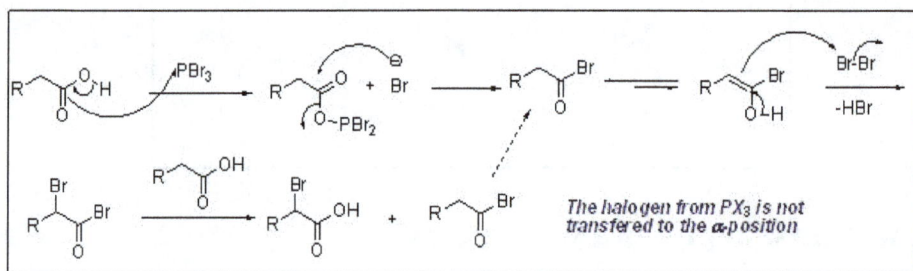

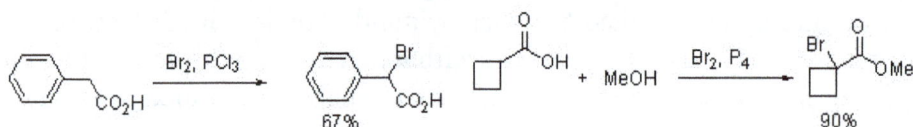

The amino group may also be introduced by Gabriel procedure (4.2.2) which gives better yield compared to that of the above described reaction with ammonia as an aminating agent.

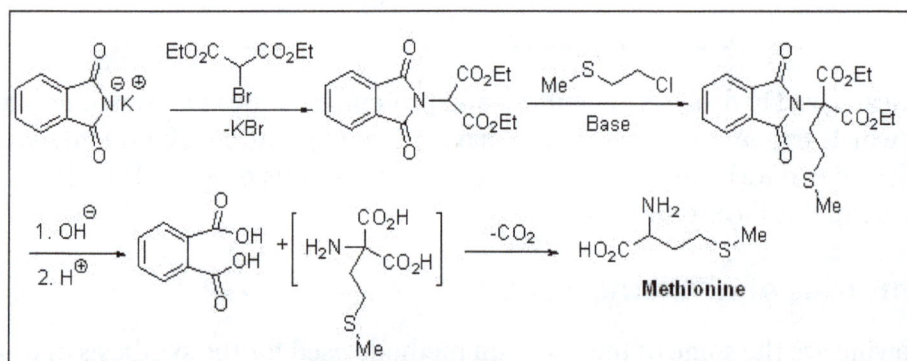

The amino group can also be introduced via nitrosation followed by reduction and hydrolysis processes.

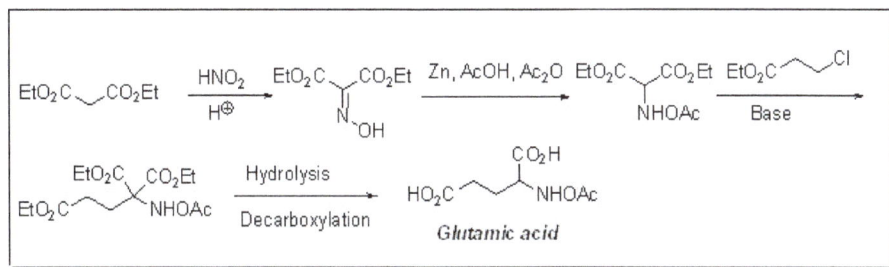

The Strecker Synthesis

The condensation of aldehydes with amine gives imine that can undergo reaction with cyanide ion in situ to give an α -aminonitrile which on hydrolysis gives α -amino acid.

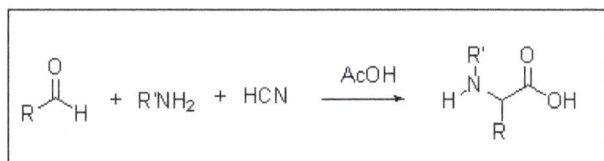

Bucherer-Bergs Reaction

Ketones proceed reaction with ammonium carbonate in presence of cyanide ion to afford hydantoin that could be hydrolyzed to α -amino acid.

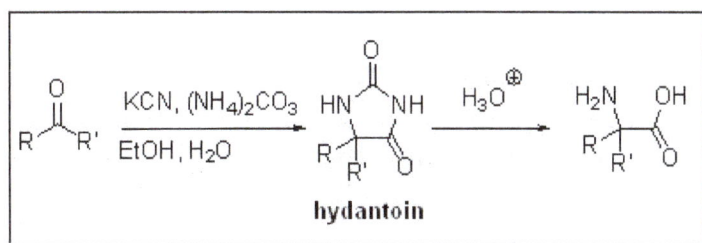

Mechanism

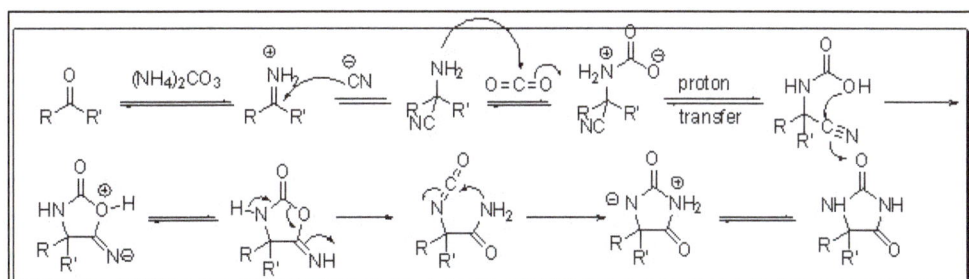

Examples

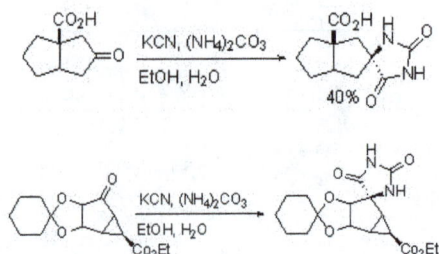

The Synthesis of Peptides

In the synthesis of peptides one amino acid is protected at its amino end with group Y and the second is protected at its carboxyl end with a group Z. The condensation of these protected amino acid derivatives is then carried out using dehydrating agent such as DCC to generate peptide bond. As per the requirement, one of the protecting groups is now removed and a third protected amino acid is introduced to the second peptide bond. Repetition of the procedure affords the desired peptide.

Protection and Deprotection of Carboxyl Group

- The carboxyl group is normally protected by converting it into its t -butyl ester using isobutylene in the presence of sulfuric acid.

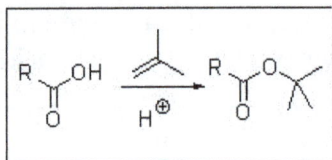

- The protecting group could also be easily removed using mild acid hydrolysis via the formation of a t-butyl carbocation.

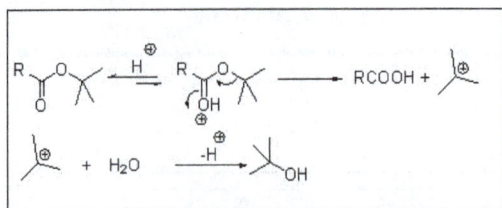

Protection and Deprotection of Amino Group

- (9-Fluorenyl)methyoxycarbonyl group (Fmoc) is commonly used as protecting group for the protection of the amino group of amino acid which can be facilitated using Na_2CO_3 in a mixture of DME and water.

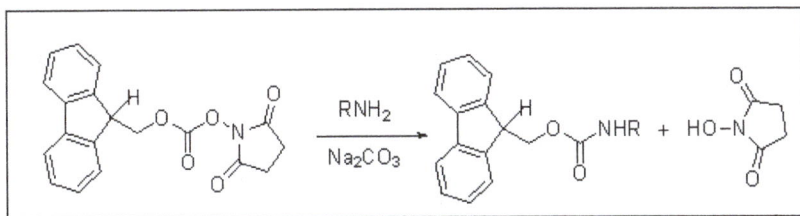

- The important characteristic of this protecting group is that it can be easily re-moved by treatment with amine base, such as piperidine.

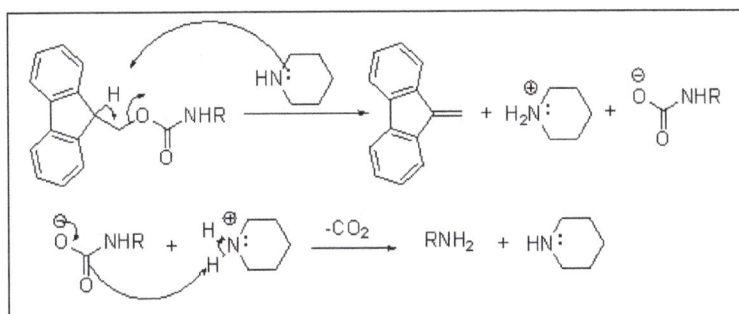

Synthesis of Tripeptide

Let us try the synthesis a tripeptide having Phe-Gly-Ala sequence. Coupling of Fmoc-gly-cine to the free amino group of t-butyl protected alanine could be effected by the reagent 1,3-dicyclohexylcarbodiimide (DCC) in the presence of N-hyroxysuccinimide (NHS). DCC is converted into DCU. The resulting protected dipipetide could be deproteced using piperdine and coupled with Fmoc-phenylalanine to give the protected tripeptide Fmoc-Phe-Gly-Ala- t Bu. The protecting groups could be deprotected using weak base (Fmoc) and mild acid (t Bu) to afford the target peptide, Phe-Gly-Ala.

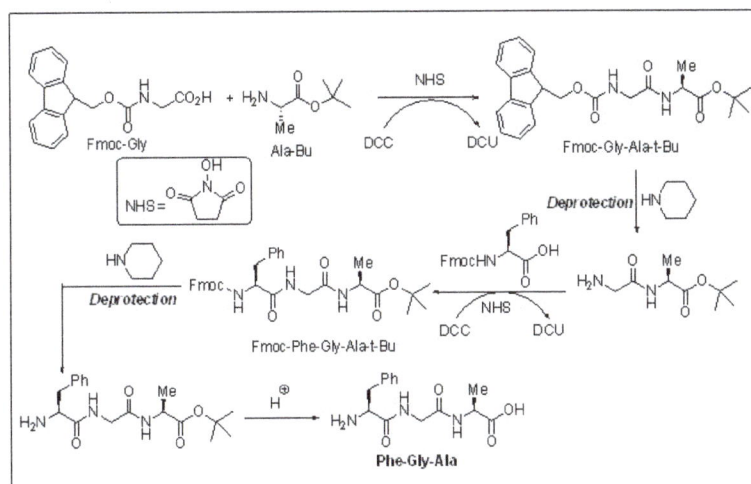

The Role of DCC and NHS in Peptide Synthesis

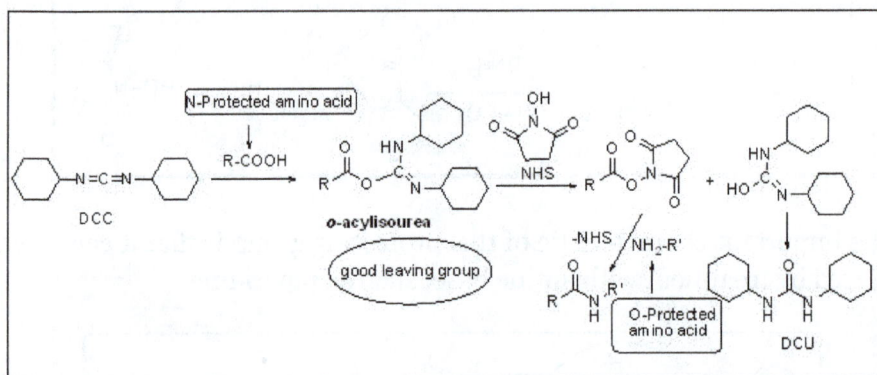

References

- Osby, J. O.; Martin, M. G.; Ganem, B. (1984). "An Exceptionally Mild Deprotection of Phthalim-ides". Tetrahedron Lett. 25 (20): 2093. doi:10.1016/S0040-4039(01)81169-2

- Ritter, John J.; Minieri, P. Paul (1948). "A New Reaction of Nitriles. I. Amides from Alkenes and Mononitriles". Journal of the American Chemical Society. 70 (12): 4045–8. PMID 18105932. doi:10.1021/ja01192a022

- Krimen, L.I.; Cota, Donald J. (1969). Adams, Rodger, ed. Organic Reaction Volume 17. London: John Wiley & Sons, Inc. pp. 213–326. ISBN 9780471196150. doi:10.1002/0471264180.or017.03

- Zil'berman, E. N. (1960). "Some reactions of nitriles with the formation of a new nitro-gen–carbon bond". Russian Chemical Reviews. 29 (6): 331–340. doi:10.1070/RC1960v-029n06ABEH001235

- Ritter, John J.; Kalish, Joseph (1948). "A New Reaction of Nitriles. II. Synthesis of t-Carbin-amines". Journal of the American Chemical Society. 70 (12): 4048–50. PMID 18105933. doi:10.1021/ja01192a023

- Rappoport, Zvi; Meyers, A. I.; Sircar, J. C. (1970). The Cyano Group (1st ed.). Charlottesville, VA: Wiley Interscience. pp. 341–421. ISBN 9780471709138. doi:10.1002/9780470771242.ch8

- Ritter, J.J.; Kalish, J. (1964). "α,α-Dimethyl-β-phenethylamine". Organic Syntheses. 42: 16. doi:10.15227/orgsyn.042.0016

- Ritter, JJ; Minieri, PP (1948). "A New Reaction of Nitriles. I. Amides from Alkenes and Mononi-triles". Journal of the American Chemical Society. 70 (12): 4045–8. PMID 18105932. doi:10.1021/ja01192a022

- Bishop, Roger (1991). "Section 1.9 – Ritter-type Reactions". Comprehensive Organic Synthe-sis Volume 6: Heteroatom Manipulation. Comprehensive Organic Synthesis. pp. 261–300. ISBN 9780080359298. doi:10.1016/B978-0-08-052349-1.00159-1

- Mattes, Susan L.; Farid, Samir (1980). "Photosensitized electron-transfer reactions of pheny-lacetylene". Journal of the Chemical Society, Chemical Communications (3): 126. doi:10.1039/C39800000126

- Sheehan, J. C.; Bolhofer, V. A. (1950). "An Improved Procedure for the Condensation of Po-tassium Phthalimide with Organic Halides". J. Am. Chem. Soc. 72 (6): 2786. doi:10.1021/ja01162a527

- Clarke, H. T.; Gillespie, H. B.; Weisshaus, S. Z. (1933). "The Action of Formaldehyde on Amines and Amino Acids". Journal of the American Chemical Society. 55 (11): 4571. doi:10.1021/ja01338a041

- Gibson, M.S.; Bradshaw, R.W. (1968). "The Gabriel Synthesis of Primary Amines". Angew. Chem. Int. Ed. Engl. 7 (12): 919. doi:10.1002/anie.196809191

- Asymmetric Catalysis of the Strecker Amino Acid Synthesis by a Cyclic Dipeptide Mani S. Iyer,, Kenneth M. Gigstad, Nivedita D. Namdev, and Mark Lipton Journal of the American Chemical Society 1996 118 (20), 4910–4911 doi:10.1021/ja952686e

An Integrated Study of Molecular Rearrangements

When the carbon skeleton of a molecule has been rearranged to give a structural isomer of the original molecule, a molecular rearrangement is said to have taken place. The major components of organic synthesis are discussed in this chapter.

Rearrangement Reaction

A rearrangement reaction is a broad class of organic reactions where the carbon skeleton of a molecule is rearranged to give a structural isomer of the original molecule. Often a substituent moves from one atom to another atom in the same molecule. In the example below the substituent R moves from carbon atom 1 to carbon atom 2:

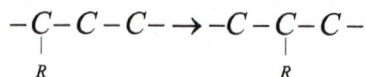

$$-\underset{\underset{R}{|}}{C}-C-C- \rightarrow -C-\underset{\underset{R}{|}}{C}-C-$$

Intermolecular rearrangements also take place.

A rearrangement is not well represented by simple and discrete electron transfers (represented by curly arrows in organic chemistry texts). The actual mechanism of alkyl groups moving, as in Wagner-Meerwein rearrangement, probably involves transfer of the moving alkyl group fluidly along a bond, not ionic bond-breaking and forming. In pericyclic reactions, explanation by orbital interactions give a better picture than simple discrete electron transfers. It is, nevertheless, possible to draw the curved arrows for a sequence of discrete electron transfers that give the same result as a rearrangement reaction, although these are not necessarily realistic. In allylic rearrangement, the reaction is indeed ionic.

Three key rearrangement reactions are 1,2-rearrangements, pericyclic reactions and olefin metathesis.

1,2-rearrangements

A 1,2-rearrangement is an organic reaction where a substituent moves from one atom to another atom in a chemical compound. In a 1,2 shift the movement involves two adjacent atoms but moves over larger distances are possible. Examples are the Wagner-Meerwein rearrangement:

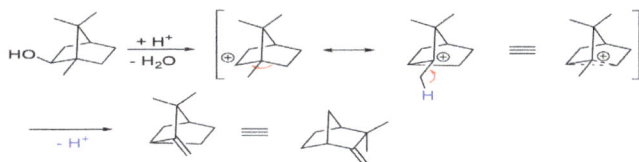

and the Beckmann rearrangement:

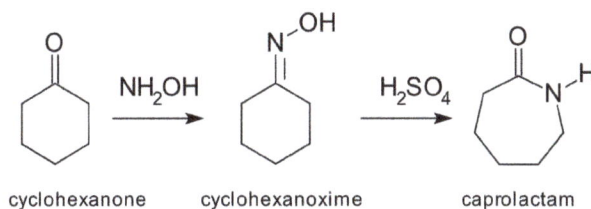

cyclohexanone   cyclohexanoxime   caprolactam

Pericyclic Reactions

A pericyclic reaction is a type of reaction with multiple carbon-carbon bond making and breaking wherein the transition state of the molecule has a cyclic geometry, and the reaction progresses in a concerted fashion. Examples are hydride shifts

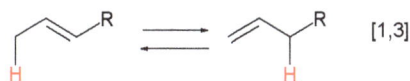

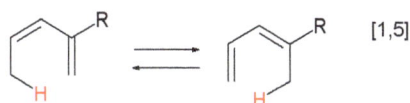

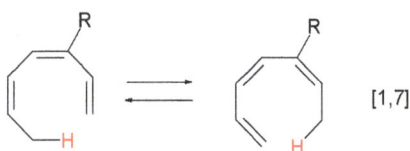

[1,3]

[1,5]

[1,7]

and the Claisen rearrangement:

Olefin Metathesis

Olefin metathesis is a formal exchange of the alkylidene fragments in two alkenes. It is a catalytic reaction with carbene, or more accurately, transition metal carbene complex intermediates.

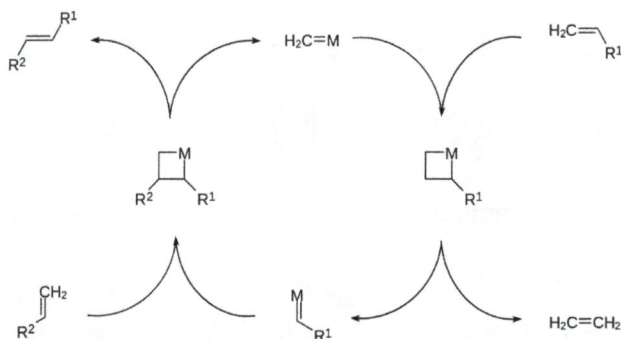

In this example, a vinyl compound is dimerized with the expulsion of ethene.

1,2-rearrangement

A 1,2-rearrangement or 1,2-migration or 1,2-shift or Whitmore 1,2-shift is an organic reaction where a substituent moves from one atom to another atom in a chemical compound. In a 1,2 shift the movement involves two adjacent atoms but moves over larger distances are possible. In the example below the substituent R moves from carbon atom C^2 to C^3.

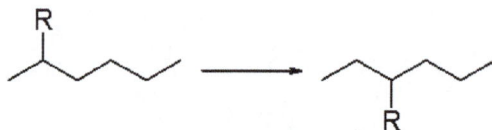

The rearrangement is intramolecular and the starting compound and reaction product are structural isomers. The 1,2-rearrangement belongs to a broad class of chemical reactions called rearrangement reactions.

A rearrangement involving a hydrogen atom is called a 1,2-hydride shift. If the substituent being rearranged is an alkyl group, it is named according to the alkyl group's anion: i.e. 1,2-methanide shift, 1,2-ethanide shift, etc.

Reaction Mechanism

A 1,2-rearrangement is often initialised by the formation of a reactive intermediate such as:

- a carbocation by heterolysis in a nucleophilic rearrangement or anionotropic rearrangement

- a carbanion in a electrophilic rearrangement or cationotropic rearrangement

- a free radical by homolysis

- a nitrene.

The driving force for the actual migration of a substituent in step two of the rearrangement is the formation of a more stable intermediate. For instance a tertiary carbocation is more stable than a secondary carbocation and therefore the S_N1 reaction of neopentyl bromide with ethanol yields tert-pentyl ethyl ether.

Carbocation rearrangements are more common than the carbanion or radical counterparts. This observation can be explained on the basis of Hückel's rule. A cyclic carbocationic transition state is aromatic and stabilized because it holds 2 electrons. In an anionic transition state on the other hand 4 electrons are present thus antiaromatic and destabilized. A radical transition state is neither stabilized or destabilized.

The most important carbocation 1,2-shift is the Wagner–Meerwein rearrangement. A carbanionic 1,2-shift is involved in the benzilic acid rearrangement.

Radical 1,2-rearrangements

The first radical 1,2-rearrangement reported by Heinrich Otto Wieland in 1911 was the conversion of bis(triphenylmethyl)peroxide 1 to the tetraphenylethane 2.

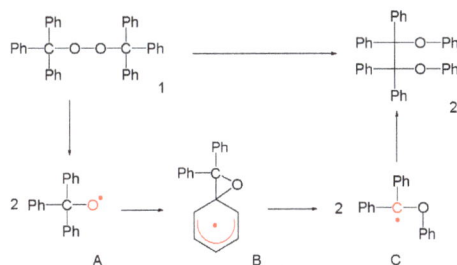

The reaction proceeds through the triphenylmethoxyl radical A, a rearrangement to diphenylphenoxymethyl C and its dimerization. It is unclear to this day whether in this rearrangement the cyclohexadienyl radical intermediate B is a transition state or a reactive intermediate as it (or any other such species) has thus far eluded detection by ESR spectroscopy.

An example of a less common radical 1,2-shift can be found in the gas phase pyrolysis of certain polycyclic aromatic compounds. The energy required in an aryl radical for the 1,2-shift can be high (up to 60 kcal/mol or 250 kJ/mol) but much less than that required for a proton abstraction to an aryne (82 kcal/mol or 340 kJ/mol). In alkene radicals proton abstraction to an alkyne is preferred.

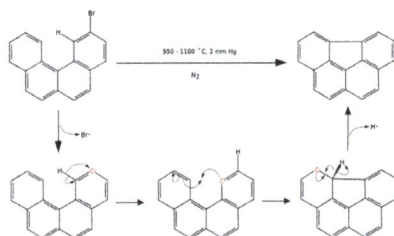

1,2 Rearrangements

The following mechanisms involve a 1,2-rearrangement:

- 1,2-Wittig rearrangement
- Alpha-ketol rearrangement
- Beckmann rearrangement
- Benzilic acid rearrangement
- Brook rearrangement
- Criegee rearrangement
- Curtius rearrangement
- Dowd–Beckwith ring expansion reaction
- Favorskii rearrangement
- Friedel–Crafts reaction
- Fritsch–Buttenberg–Wiechell rearrangement
- Halogen dance rearrangement
- Hofmann rearrangement
- Lossen rearrangement
- Pinacol rearrangement
- Seyferth–Gilbert homologation
- S_N1 reaction (generally)
- Stevens rearrangement
- Wagner–Meerwein rearrangement
- Westphalen–Lettré rearrangement
- Wolff rearrangement

1,3-rearrangements

1,3-rearrangements take place over 3 carbon atoms. Examples:

- the Fries rearrangement
- a 1,3-alkyl shift of verbenone to chrysanthenone

Pericyclic Reaction

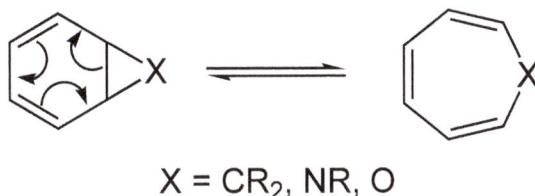

$$X = CR_2, NR, O$$

In organic chemistry, a pericyclic reaction is a type of organic reaction wherein the transition state of the molecule has a cyclic geometry, and the reaction progresses in a concerted fashion. Pericyclic reactions are usually rearrangement reactions. The major classes of pericyclic reactions are:

Name	Bond changes	
	Sigma	Pi
Electrocyclic reaction	+ 1	-1
Cycloaddition	+2	-2
Sigmatropic reaction	0	0
Group transfer reaction	+ 1	-1
Cheletropic reaction	+ 2	- 2
Dyotropic reaction	0	0

In general, these are considered to be equilibrium processes, although it is possible to push the reaction in one direction by designing a reaction by which the product is at a significantly lower energy level; this is due to a unimolecular interpretation of Le Chatelier's principle. Pericyclic reactions often have related stepwise radical processes associated with them. Some pericyclic reactions, such as the [2+2] cycloaddition, are 'controversial' because their mechanism is not definitively known to be concerted (or may depend on the reactive system). Pericyclic reactions also often have metal-catalyzed analogs, although usually these are also not technically pericyclic, since they proceed via metal-stabilized intermediates, and therefore are not concerted.

A large photoinduced hydrogen sigmatropic shift was utilized in a corrin synthesis performed by Albert Eschenmoser containing a 16π system.

Due to the principle of microscopic reversibility, there is a parallel set of "retro" pericyclic reactions, which perform the reverse reaction.

Pericyclic Reactions in Stereochemistry

It is well established that the diene can only enter in a cycloaddition reaction with a dienophile in the cisoid form and the rate of the reaction with the open chain dienes depends on the equilibrium proportions of the cisoid/transoid conformers. Thus substituents in the diene can significantly affect the rate of the reaction not only by their electronic character but by their influence on the relative proportions of the different conformers.

Thus for example cis I -substituted butadiene I is less reactive than its trans isomer II since a bulky R disfavors the cisoid conformation. Bulky 2-substituents in the diene favor the cisoid conformation more than the transoid and thus the diene in this case is more reactive.

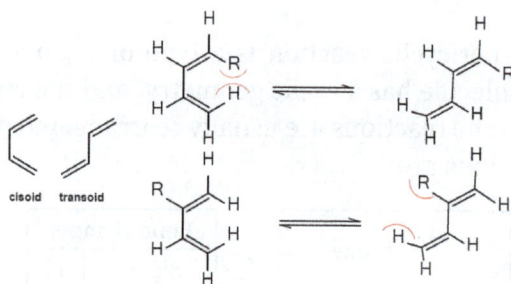

Pericyclic Reactions in Biochemistry

Pericyclic reactions also occur in several biological processes:

- Claisen rearrangement of chorismate to prephenate in almost all prototrophic organisms

- [1,5]-sigmatropic shift in the transformation of precorrin-8x to hydrogenoby-rinic acid

- non-enzymatic, photochemical electrocyclic ring opening and a (1,7) sigma-tropic hydride shift in vitamin D synthesis

- Isochorismate pyruvate lyase catalyzed conversion of Isochorismate into salic-ylate and Pyruvate

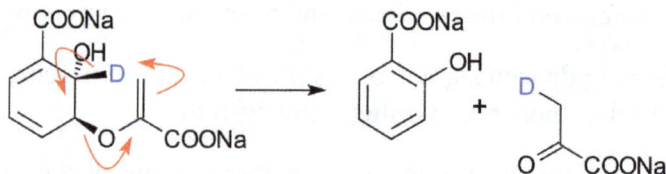

Olefin Metathesis

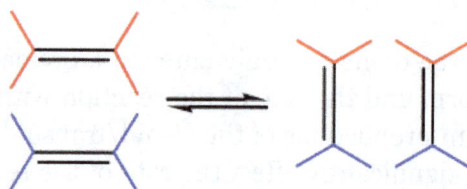

Olefin metathesis is an organic reaction that entails the redistribution of fragments of alkenes (olefins) by the scission and regeneration of carbon-carbon double bonds. Catalysts for this reaction have evolved rapidly for the past few decades. Because of the relative simplicity of olefin metathesis, it often creates fewer undesired by-products and hazardous wastes than alternative organic reactions. For their elucidation of the reaction mechanism and their discovery of a variety of highly efficient and selective catalysts, Yves Chauvin, Robert H. Grubbs, and Richard R. Schrock were collectively awarded the 2005 Nobel Prize in Chemistry.

Catalysts

The reaction is catalyzed by metal complexes. Traditional catalysts are prepared by a reaction of the metal halides with alkylation agents, for example WCl_6–$EtOH$–$EtAlCl_2$. The traditional, industrial catalysts are ill-defined and used mainly for Petroleum products. Modern catalysts are well-defined organometallic compounds that come in two main categories, commonly known as Schrock catalysts and Grubbs' catalysts. Schrock catalysts are molybdenum(IV)- and tungsten(IV)-based, and are examples of Schrock alkylidenes.

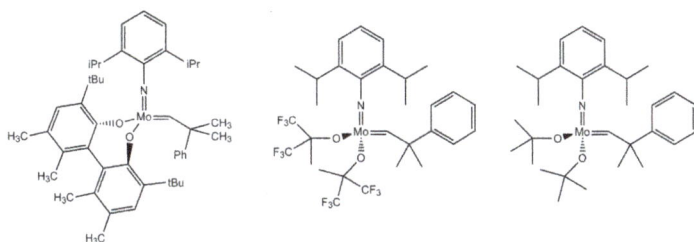

Grubbs' catalysts, on the other hand, are ruthenium(II) carbenoid complexes. Grubbs' catalysts are often modified with a chelating isopropoxystyrene ligand to form the related Hoveyda–Grubbs catalyst.

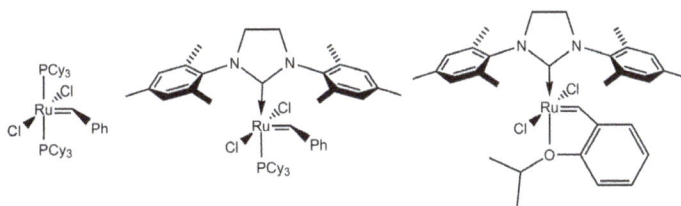

Applications

Olefin metathesis was first commercialized in petroleum reformation for the synthesis of higher olefins from the products (alpha-olefins) from the Shell higher olefin process (SHOP) under high pressure and high temperatures.

Modern catalysts can be used for a variety of specialized organic compounds and monomers. Modern applications include the synthesis of pharmaceutical drugs, macrocyclic crownophanes the manufacturing of high-strength materials, the production of propylene, the preparation of cancer-targeting nanoparticles, and the conversion of renewable plant-based feedstocks into hair and skin care products.

Types of Olefin Metathesis Processes

Some important classes of olefin metathesis include:

- Cross metathesis (CM)

- Ring-opening metathesis (ROM)

- Ring-closing metathesis (RCM)

- Ring-opening metathesis polymerisation (ROMP)

- Acyclic diene metathesis (ADMET)

- Ethenolysis

Reaction Mechanism

Hérisson and Chauvin first proposed the widely accepted mechanism of transition metal alkene metathesis. The direct [2+2] cycloaddition of two alkenes is formally symmetry forbidden and thus has a high activation energy. The Chauvin mechanism involves the [2+2] cycloaddition of an alkene double bond to a transition metal alkylidene to form a metallacyclobutane intermediate. The metallacyclobutane produced can then cyclorevert to give either the original species or a new alkene and alkylidene. Interaction with the d-orbitals on the metal catalyst lowers the activation energy enough that the reaction can proceed rapidly at modest temperatures.

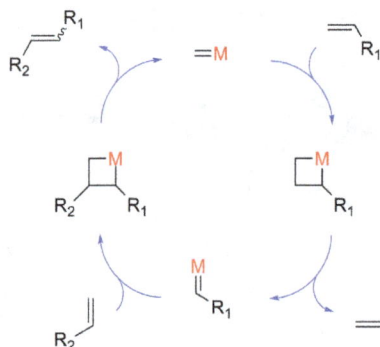

Like most chemical reactions, the metathesis pathway is driven by a thermodynamic imperative; that is, the final products are determined by the energetics of the possible products, with a distribution of products proportional to the exponential of their respective energy values. In olefin metathesis, however, this is especially relevant since all the possible products have similar energy values (all of them contain an olefin).

Because of this the product mixture can be tuned by reaction conditions, such as gas pressure and substrate concentration. In some cases a given reaction can be run in either direction to near completion.

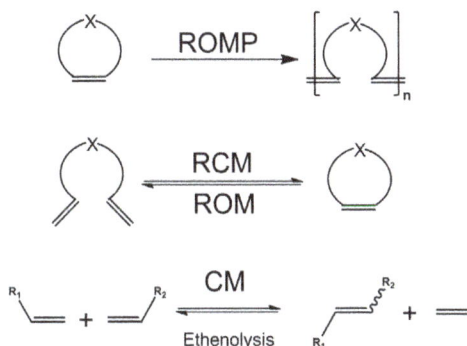

Cross metathesis and Ring-closing metathesis are often driven by the entropically favored evolution of ethylene or propylene, which are both gases. Because of this CM and RCM reactions often use alpha-olefins. The reverse reaction of CM of two alpha-olefins, ethenolysis, can be favored but requires high pressures of ethylene to increase ethylene concentration in solution. The reverse reaction of RCM, ring-opening metathesis, can likewise be favored by a large excess of an alpha-olefin, often styrene. Ring opening metathesis usually involves a strained alkene (often a norbornene) and the release of ring strain drives the reaction. Ring-closing metathesis, conversely, usually involves the formation of a five- or six-membered ring, which is energetically favorable; although these reactions tend to also evolve ethylene, as previously discussed. RCM has been used to close larger macrocycles, in which case the reaction may be kinetically controlled by running the reaction at extreme dilutions. The same substrates that undergo RCM can undergo acyclic diene metathesis, with ADMET favored at high concentrations. The Thorpe–Ingold effect may also be exploited to improve both reaction rates and product selectivity.

Cross-metathesis is synthetically equivalent to (and has replaced) a procedure of ozonolysis of an alkene to two ketone fragments followed by the reaction of one of them with a Wittig reagent.

Historical Overview

Known chemistry prior to the advent of olefin metathesis was introduced in the 1950s by Karl Ziegler, who as part of ongoing work in what would later become known as Ziegler–Natta catalysis studied ethylene polymerization, which on addition of certain metals resulted in 1-butene instead of a saturated long-chain hydrocarbon.

In 1960 a Du Pont research group polymerized norbornene to polynorbornene using *lithium aluminum tetraheptyl* and titanium tetrachloride (a patent by this company on this topic dates back to 1955),

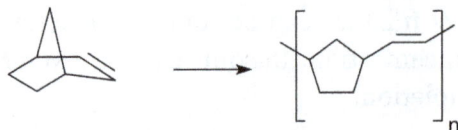

a reaction then classified as a so-called coordination polymerization. According to the then proposed reaction mechanism a RTiX titanium intermediate first coordinates to the double bond in a pi complex. The second step then is a concerted SNi reaction breaking a CC bond and forming a new alkylidene-titanium bond; the process then repeats itself with a second monomer:

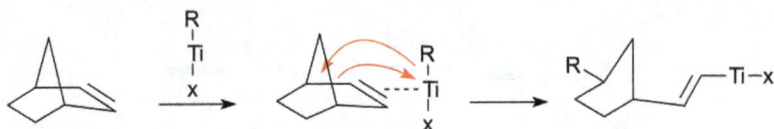

Only much later the polynorbornene was going to be produced through ring opening metathesis polymerisation. The DuPont work was led by Herbert S. Eleuterio. Giulio Natta in 1964 also observed the formation of an unsaturated polymer when polymerizing cyclopentene with tungsten and molybdenum halides.

In a third development leading up to olefin metathesis, researchers at Phillips Petroleum Company in 1964 described olefin disproportionation with catalysts molybdenum hexacarbonyl, tungsten hexacarbonyl, and molybdenum oxide supported on alumina for example converting propylene to an equal mixture of ethylene and 2-butene for which they proposed a reaction mechanism involving a cyclobutane (they called it a quasicyclobutane) – metal complex:

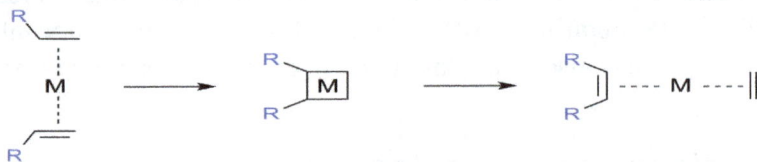

This particular mechanism is symmetry forbidden based on the Woodward–Hoffmann rules first formulated two years earlier. Cyclobutanes have also never been identified in metathesis reactions another reason why it was quickly abandoned.

Then in 1967 researchers at the Goodyear Tire and Rubber Company described a novel catalyst system for the metathesis of 2-pentene based on tungsten hexachloride, ethanol the organoaluminum compound EtAlMe$_2$ and also proposed a name for this reaction type: olefin metathesis. Formerly the reaction had been called "olefin disproportionation."

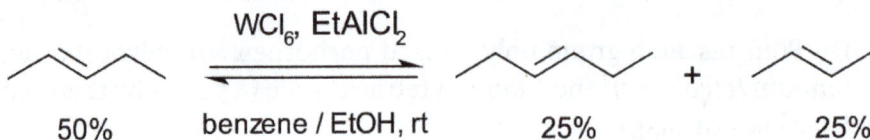

In this reaction 2-pentene forms a rapid (a matter of seconds) chemical equilibrium with 2-butene and 3-hexene. No double bond migrations are observed; the reaction can be started with the butene and hexene as well and the reaction can be stopped by addition of methanol.

The Goodyear group demonstrated that the reaction of regular 2-butene with its all-deuterated isotopologue yielded $C_4H_4D_4$ with deuterium evenly distributed. In this way they were able to differentiate between a transalkylidenation mechanism and a transalkylation mechanism (ruled out):

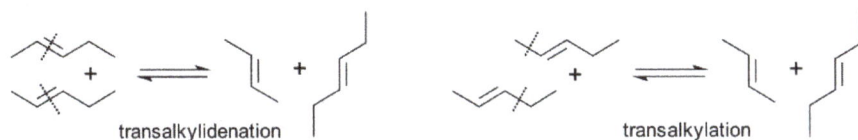

In 1971 Chauvin proposed a four-membered metallacycle intermediate to explain the statistical distribution of products found in certain metathesis reactions. This mechanism is today considered the actual mechanism taking place in olefin metathesis.

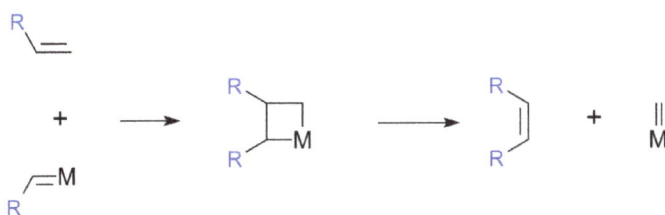

The active catalyst, a metallocarbene, was discovered by in 1964 by E. O. Fischer. Chauvin's experimental evidence was based on the reaction of cyclopentene and 2-pentene with the homogeneous catalyst tungsten(VI) oxytetrachloride and tetrabutyltin:

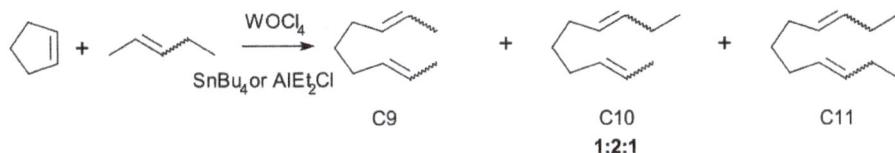

The three principal products C9, C10 and C11 are found in a 1:2:1 regardless of conversion. The same ratio is found with the higher oligomers. Chauvin also explained how the carbene forms in the first place: by alpha-hydride elimination from a carbon metal single bond. For example, propylene (C3) forms in a reaction of 2-butene (C4) with tungsten hexachloride and tetramethyltin (C1).

In the same year Pettit who synthesised cyclobutadiene a few years earlier independently came up with a competing mechanism. It consisted of a tetramethylene intermediate with sp³ hybridized carbon atoms linked to a central metal atom with multiple three-center two-electron bonds.

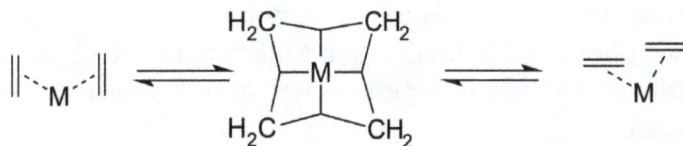

Experimental support offered by Pettit for this mechanism was based on an observed reaction inhibition by carbon monoxide in certain metathesis reactions of 4-nonene with a tungsten metal carbonyl

Robert H. Grubbs got involved in metathesis in 1972 and also proposed a metallacycle intermediate but one with four carbon atoms in the ring. The group he worked in reacted 1,4-dilithiobutane with tungsten hexachloride in an attempt to directly produce a cyclomethylenemetallacycle producing an intermediate, which yielded products identical with those produced by the intermediate in the olefin metathesis reaction. This mechanism is pairwise:

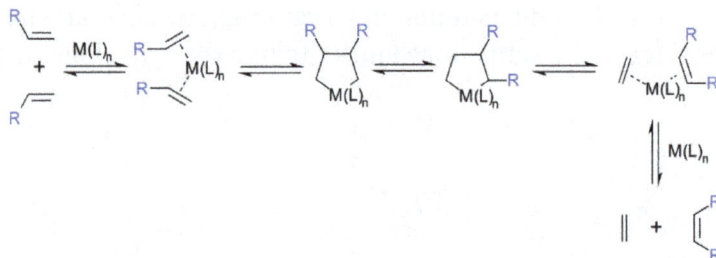

In 1973 Grubbs found further evidence for this mechanism by isolating one such metallacycle not with tungsten but with platinum by reaction of the dilithiobutane with *cis-bis(triphenylphosphine)dichloroplatinum(II)*

In 1975 Katz also arrived at a metallacyclobutane intermediate consistent with the one proposed by Chauvin He reacted a mixture of cyclooctene, 2-butene and 4-octene with a molybdenum catalyst and observed that the unsymmetrical C14 hydrocarbon reaction product is present right from the start at low conversion.

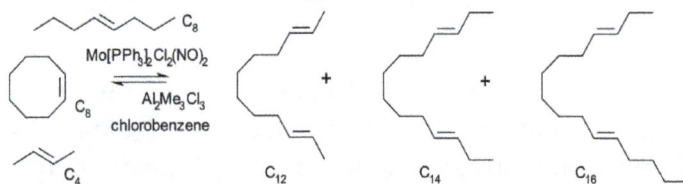

In any of the pairwise mechanisms with olefin pairing as rate-determining step this compound, a secondary reaction product of C12 with C6, would form well after formation of the two primary reaction products C12 and C16.

In 1974 Casey was the first to implement carbenes into the metathesis reaction mechanism:

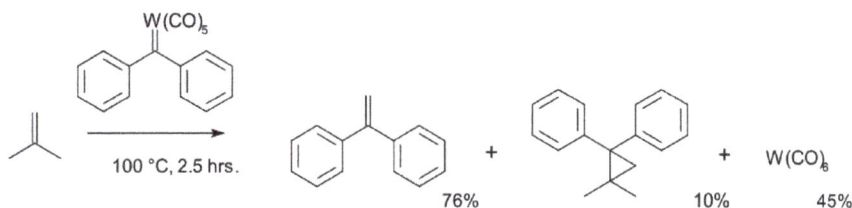

Grubbs in 1976 provided evidence against his own updated pairwise mechanism:

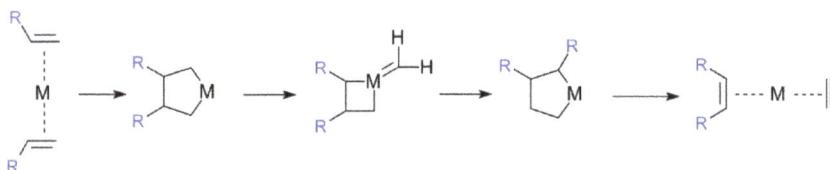

with a 5-membered cycle in another round of isotope labeling studies in favor of the 4-membered cycle Chauvin mechanism:

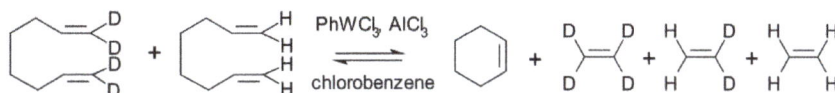

In this reaction the ethylene product distribution (d_4, d_2, d_0) at low conversion was found to be consistent with the carbene mechanism. On the other hand, Grubbs did not rule out the possibility of a tetramethylene intermediate.

The first practical metathesis system was introduced in 1978 by Tebbe based on the (what later became known as the) Tebbe reagent. In a model reaction isotopically labeled carbon atoms in isobutene and methylenecyclohexane switched places:

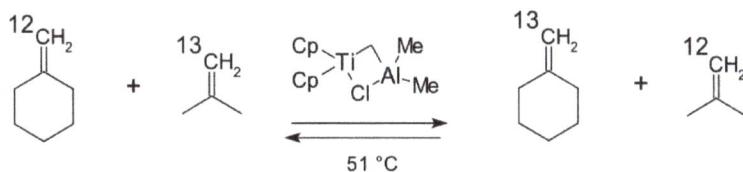

The Grubbs group then isolated the proposed metallacyclobutane intermediate in 1980 also with this reagent together with 3-methyl-1-butene:

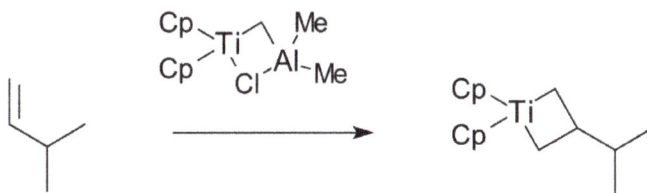

They isolated a similar compound in the total synthesis of capnellene in 1986:

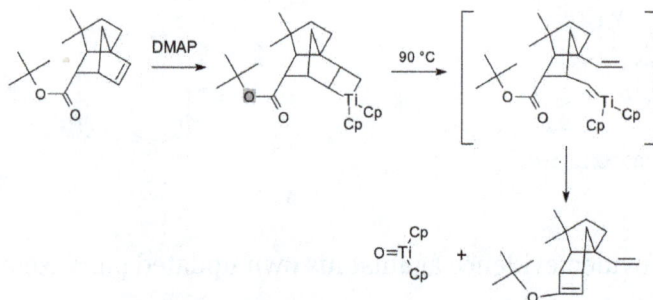

In that same year the Grubbs group proved that metathesis polymerization of norbornene by Tebbe's reagent is a living polymerization system and a year later Grubbs and Schrock co-published an article describing living polymerization with a tungsten carbene complex While Schrock focussed his research on tungsten and molybdenum catalysts for olefin metathesis, Grubbs started the development of catalysts based on ruthenium, which proved to be less sensitive to oxygen and water and therefore more functional group tolerant.

Grubbs Catalysts

In the 1960s and 1970s various groups reported the ring-opening polymerization of norbornene catalyzed by hydrated trichlorides of ruthenium and other late transition metals in polar, protic solvents. This prompted Robert H. Grubbs and coworkers to search for well-defined, functional group tolerant catalysts based on ruthenium. The Grubbs group successfully polymerized the 7-oxo norbornene derivative using ruthenium trichloride, osmium trichloride as well as tungsten alkylidenes. They identified a Ru(II) carbene as an effective metal center and in 1992 published the first well-defined, ruthenium-based olefin metathesis catalyst, $(PPh_3)_2Cl_2Ru=CHCH=CPh_2$:

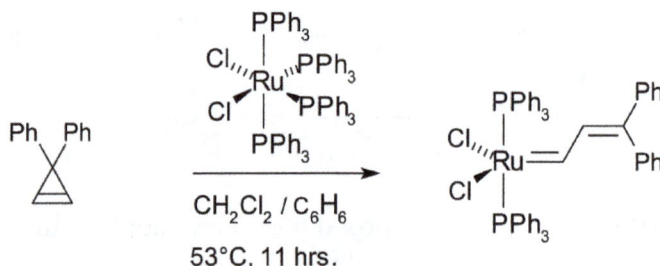

The corresponding tricyclohexylphosphine complex $(PCy_3)_2Cl_2Ru=CHCH=CPh_2$ was also shown to be active. This work culminated in the now commercially available 1st generation Grubbs catalyst.

Schrock Catalysts

Schrock entered the olefin metathesis field in 1979 as an extension of work on tantalum alkylidenes. The initial result was disappointing as reaction of CpTa(=CH-t-Bu)Cl$_2$ with ethylene yielded only a metallacyclopentane, not metathesis products:

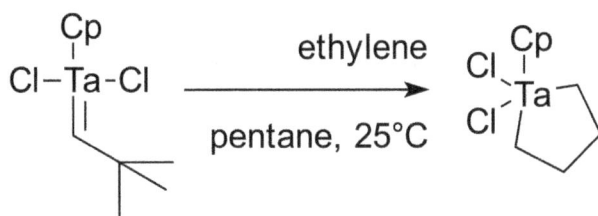

But by tweaking this structure to a PR$_3$Ta(CHt-bu)(Ot-bu)$_2$Cl (replacing chloride by t-butoxide and a cyclopentadienyl by an organophosphine, metathesis was established with cis-2-pentene. In another development, certain tungsten oxo complexes of the type W(O)(CHt-Bu)(Cl)$_2$(PEt)$_3$ were also found to be effective.

Schrock alkylidenes for olefin metathesis of the type Mo(NAr)(CHC(CH$_3$)$_2$R){OC(CH$_3$) (CF$_3$)$_2$}$_2$ were commercialized starting in 1990.

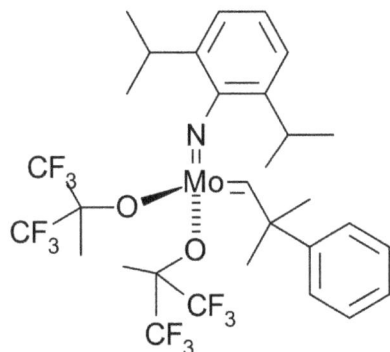

The first asymmetric catalyst followed in 1993

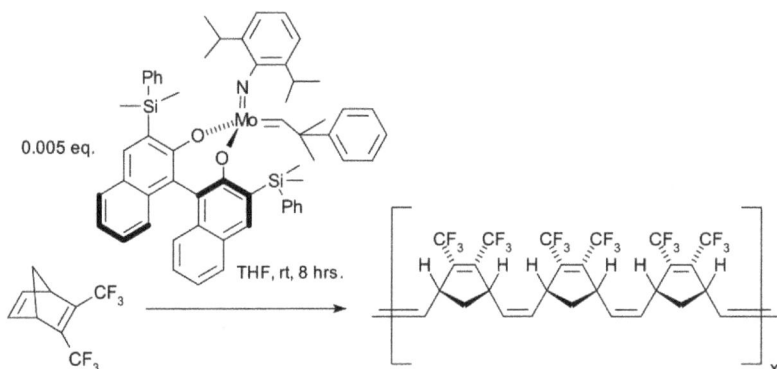

With a Schrock catalyst modified with a BINOL ligand in a norbornadiene ROMP leading to highly stereoregular cis, isotactic polymer.

Types of Rearrangements

Rearrangements are divided into intramolecular and intermolecular processes. In intramolecular process, the group that migrates is not completely detached from the system in which rearrangement is taking place. In contrast, in intermolecular process, the migrating group is first detached and later re-attached at another site.

Rearrangement to Electron Deficient Carbon

These reactions are classified according to the nature of group that migrates.

Carbon Migration

Wagner-Meerwein Rearrangement

It is one of the simplest systems where an alkyl group migrates, with its bonding pair, to an electron-deficient carbon atom.

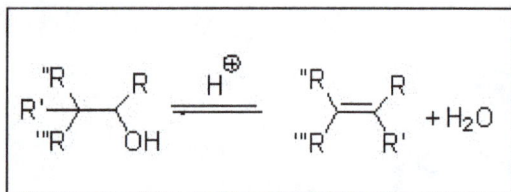

Mechanism

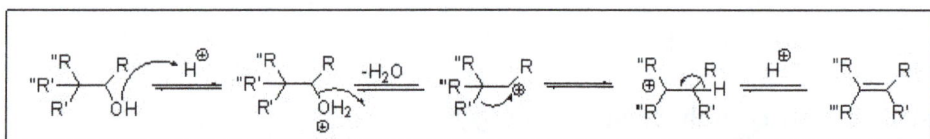

The driving force for the rearrangement resides in the greater stability of a tertiary carbocation compared to that of primary carbocation.

The classical and non-classical carbocation controversy concerned the Wagner-Meerwein rearrangement of norbornyl systems:

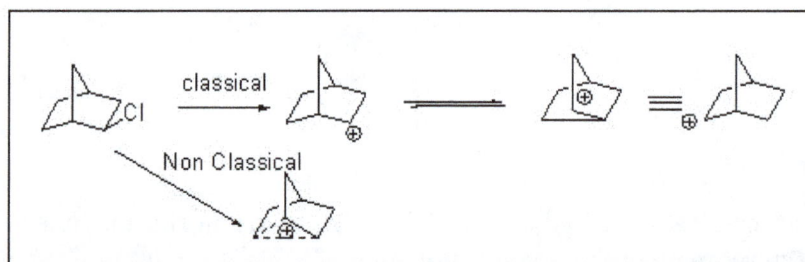

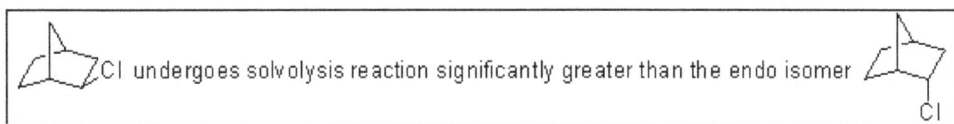

Cl undergoes solvolysis reaction significantly greater than the endo isomer

Features of this Migration

- The carbocation may be produced by a variety of ways.

- Hydrogen can also migrate in this system.

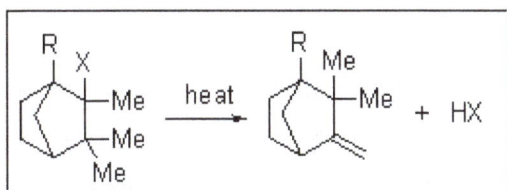

- Aryl groups have a greater migratory aptitude than alkyl group or hydrogen due to the formation of lower-energy bridged phenonium ion.

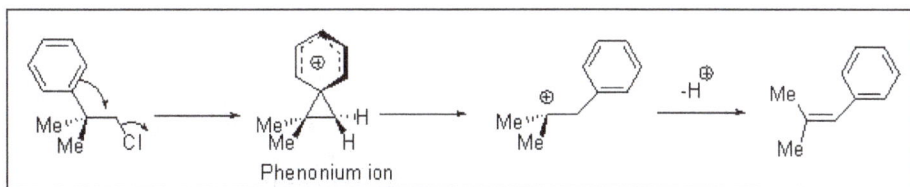

Phenonium ion

- Rearrangements in bicyclic systems are common.

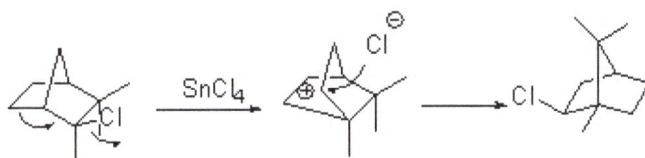

- The rearrangement is stereosepecific

- Two or more rearrangements may take place simultaneously.

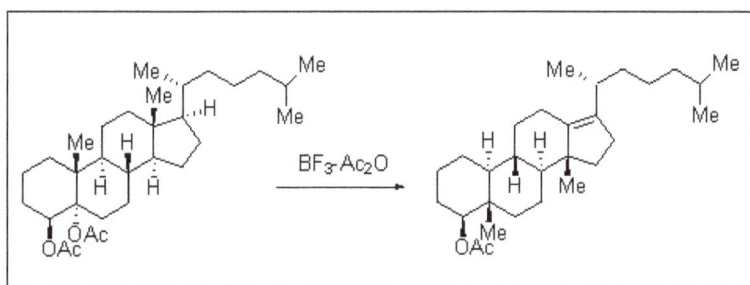

Examples

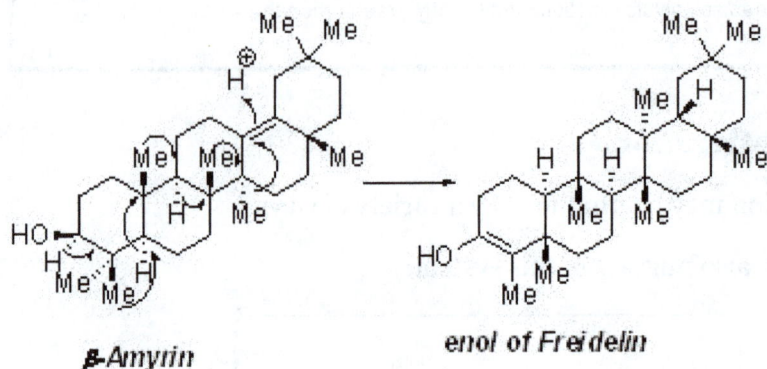

β-Amyrin

enol of Freidelin

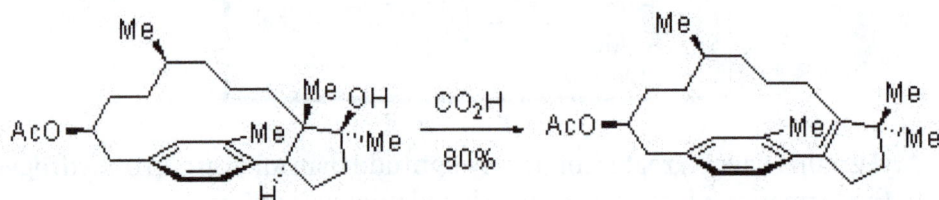

Pinacol Rearrangement

Treatment of 1,2-diols (pinacol) with acid lead to rearrangement to give ketone. Although this rearrangement fundamentally is similar to the above described Wagner-Meerwein rearrangement, but differs in that the rearranged ion, the conjugate acid of ketone, is relatively more stable than the rearranged carbocation formed in Wagner-Meerwein rearrangement. Thus, the driving force for pinacol is greater compared to Wagner-Meerwein rearrangement. However, the characteristics of the Wagner-Meerwein apply to the pinacol rearrangement.

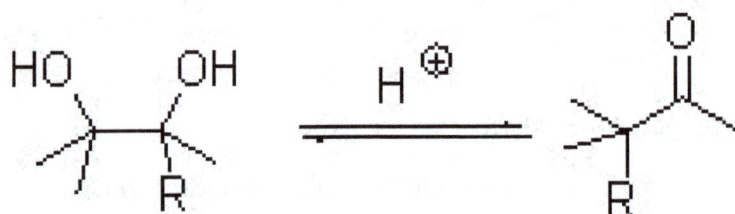

Mechanism

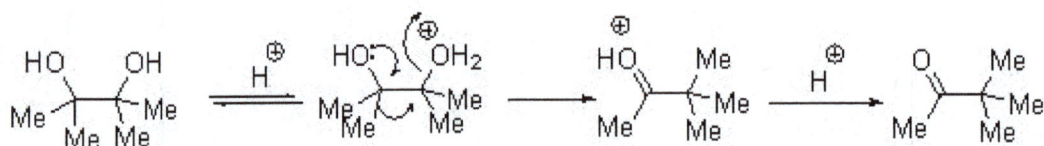

Examples

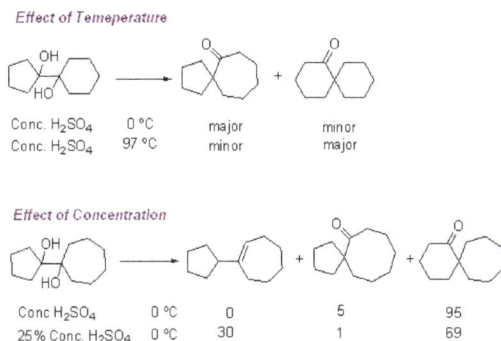

Benzilic Acid Rearrangement

1,2-Diketones that have no a -hydrogen react with hydroxide ion to give a -hydroxyacid. The best known example is the rearrangement of benzil to benzilic acid. The driving force for the reaction lies in the removel of the product by ionization of carbonyl group.

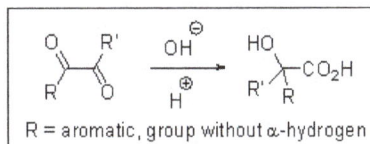

Mechanism

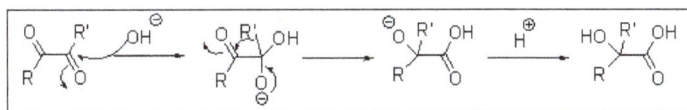

Examples

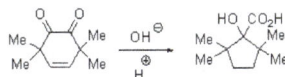

Arndt-Eistert Homologation Reaction

The reaction of acid chloride with diazomethane gives a diazoketone which is in the presence of silver oxide under heating proceeds the Wolff rearrangement to yield a ketene that is directly converted into an acid in the presence of water.

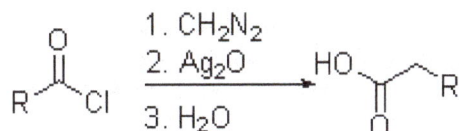

Mechanism

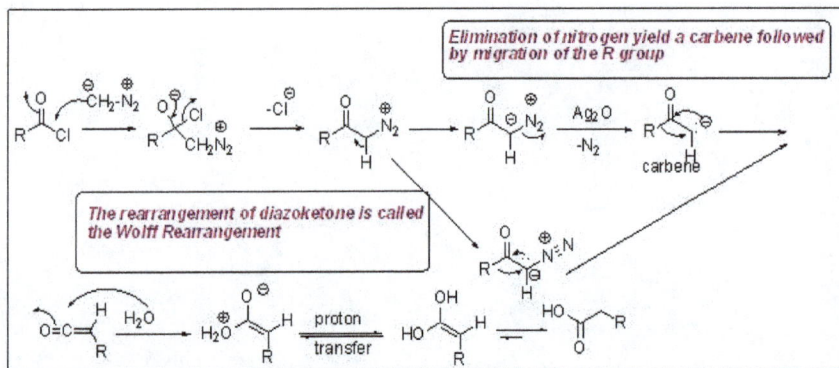

The rearrangement of diazoketone is called the Wolff Rearrangement

Elimination of nitrogen yield a carbene followed by migration of the R group

carbene

proton transfer

Examples

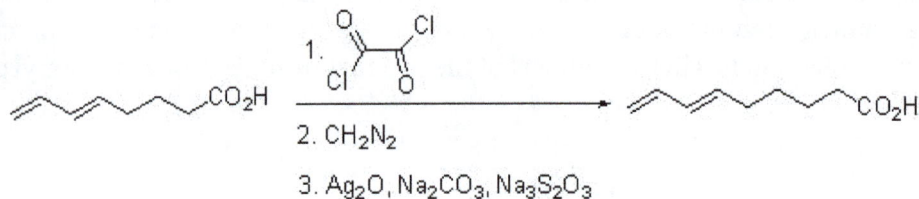

1. [reagent with Cl, O]

2. CH_2N_2

3. Ag_2O, Na_2CO_3, $Na_3S_2O_3$

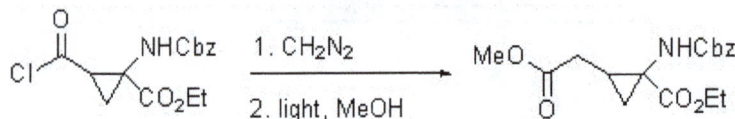

1. CH_2N_2

2. light, MeOH

Halogen, Oxygen, Sulfur, and Nitrogen Migration

In the system X-C-C-Y, an atom X with an unshared pair of electrons can assist the heterolysis of the C-Y bond. In case of unsymmetrical system, nucleophilic attack predominates at the less substituted carbon of the bridged ion that leads to rearranged skeleton.

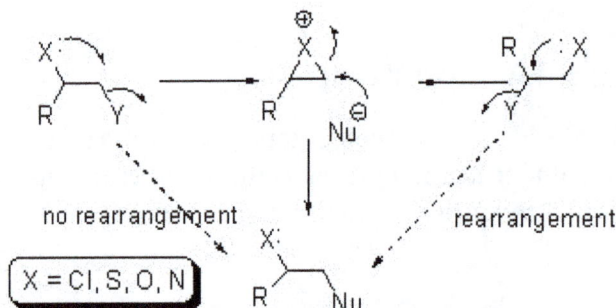

no rearrangement

rearrangement

X = Cl, S, O, N

Some Examples Follow

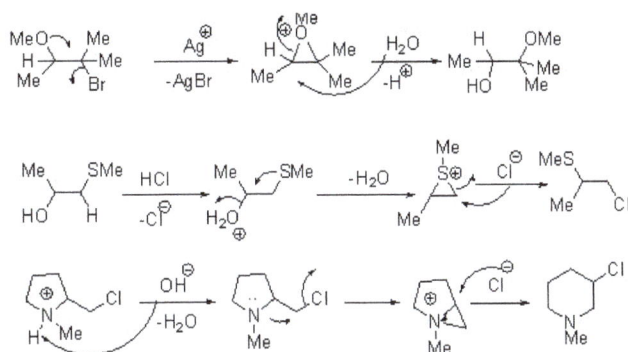

Rupe Rearrangement

The bridged cation may be produced via protonation of an unsaturated bond as in the Rupe rearrangement of a -acetylenic alcohols.

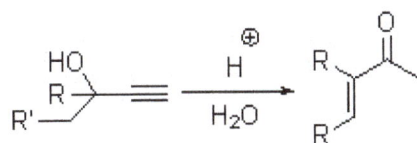

Mechanism

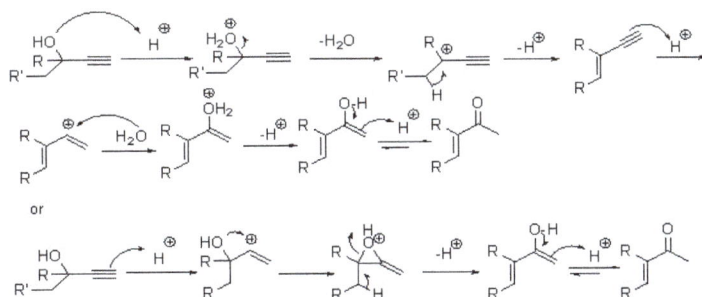

Example

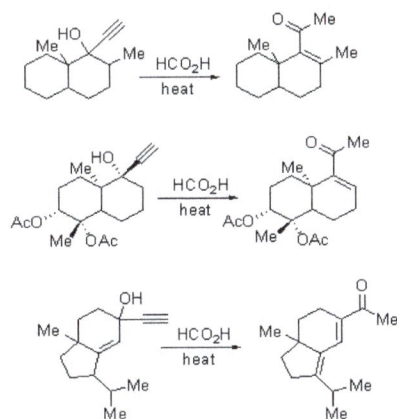

In case of a neighbouring acetoxy group, the solvolysis is assisted via a five-membered acetoxonium ion.

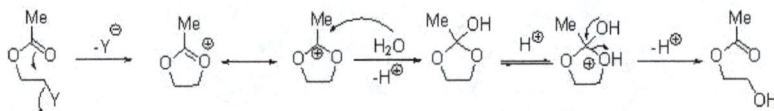

Pinacol Rearrangement

The pinacol–pinacolone rearrangement is a method for converting a 1,2-diol to a carbonyl compound in organic chemistry. The 1,2-rearrangement takes place under acidic conditions. The name of the rearrangement reaction comes from the rearrangement of pinacol to pinacolone.

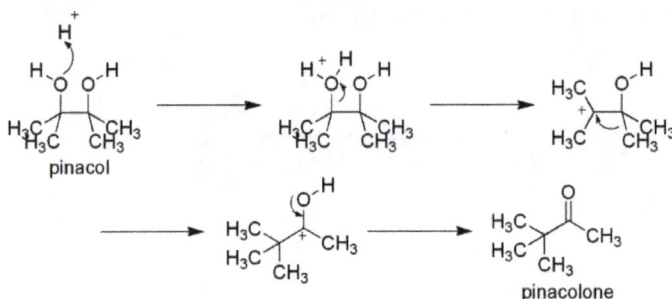

This reaction was first described by Wilhelm Rudolph Fittig in 1860 of the famed Fittig reaction involving coupling of 2 aryl halides in presence of sodium metal in dry ethereal solution.

An Overview of Mechanism(Discussion)

In the course of this organic reaction, protonation of one of the –OH groups occurs and a carbocation is formed. If both the –OH groups are not alike, then the one which yields a more stable carbocation participates in the reaction. Subsequently, an alkyl group from the adjacent carbon migrates to the carbocation center. The driving force for this rearrangement step is believed to be the relative stability of the resultant oxonium ion, which has complete octet configuration at all centers (as opposed to the preceding carbocation). The migration of alkyl groups in this reaction occurs in accordance with their usual migratory aptitude, i.e.hydride > Phenyl > tertiary carbocation (if formed by migration) > secondary carbocation (if formed by migration) > methyl cation . The conclusion which group stabilizes carbocation more effectively is migrated

Stereochemistry of the Rearrangement

In cyclic systems, the reaction presents more features of interest. In these reactions, the stereochemistry of the diol plays a crucial role in deciding the major product. An alkyl group which is situated trans- to the leaving −OH group alone may migrate. If otherwise, ring expansion occurs, i.e. the ring carbon itself migrates to the carbocation centre. This reveals another interesting feature of the reaction, viz. that it is largely concerted. There appears to be a connection between the migration origin and migration terminus throughout the reaction.

Moreover, if the migrating alkyl group has a chiral center as its key atom, the configuration at this center is *retained* even after migration takes place.

History

Although Fittig first published about the pinacol rearrangement,it was not Fittig but Aleksandr Butlerov who correctly identified the reaction products involved.

In an 1859 publication Wilhelm Rudolph Fittig described the reaction of acetone with potassium metal... Fittig wrongly assumed a molecular formula of $(C_3H_3O)_n$ for acetone, the result of a long-standing atomic weight debate finally settled at the Karlsruhe Congress in 1860. He also wrongly believed acetone to be an alcohol which he hoped to prove by forming a metal alkoxide salt. The reaction product he obtained instead he called paraceton which he believed to be an acetone dimer. In his second publication in 1860 he reacted paraceton with sulfuric acid (the actual pinacol rearrangement).

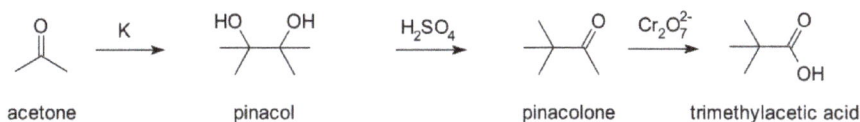

| acetone | pinacol | pinacolone | trimethylacetic acid |

Again Fittig was unable to assign a molecular structure to the reaction product which he assumed to be another isomer or a polymer. Contemporary chemists who had already adapted to the new atomic weight reality did not fare better. One of them, Charles Friedel, believed the reaction product to be the epoxide tetramethylethylene oxide in analogy with reactions of ethylene glycol. Finally Butlerov in 1873 came up with the correct structures after he independently synthesised the compound trimethylacetic acid which Friedel had obtained earlier by oxidizing with a dichromate.

Some of the problems during the determination of the structure are because carbon skeletal rearrangements were unknown at that time and therefore the new concept had to be found. Butlerov theory allowed the structure of carbon atoms in the molecule to rearrange and with this concept a structure for pinacolone could be found.

Benzilic Acid Rearrangement

The benzilic acid rearrangement is the rearrangement reaction of benzil with potassium hydroxide to benzilic acid. First performed by Justus Liebig in 1838 this reaction type is displayed by 1,2-diketones in general. The reaction product is an α-hydroxy–carboxylic acid.

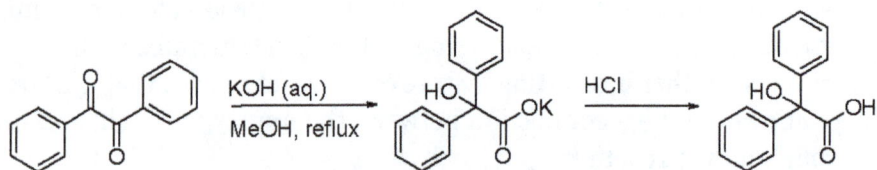

Certain acyloins also rearrange in this fashion.

This diketone reaction is related to other rearrangements: the corresponding keto-aldehyde (one alkyl group replaced by hydrogen) rearranges in a Cannizzaro reaction, the corresponding 1,2-diol reacts in a pinacol rearrangement.

Reaction Mechanism

The reaction is a representative of 1,2-rearrangements. These rearrangements usually have migrating carbocations but this reaction is unusual because it involves a migrating carbanion. The long established reaction mechanism updated with in silico data is outlined in *figure*.

A hydroxide anion attacks one of the ketone groups in 1 in a nucleophilic addition to the hydroxyl anion 2. The next step requires a bond rotation to conformer 3 which places the migrating group R in position for attack on the second carbonyl group in a concerted step with reversion of the hydroxyl group back to the carbonyl group. This sequence resembles a nucleophilic acyl substitution. Calculations show that when R is methyl the charge build-up on this group in the transition state can be as high as 0.22 and that the methyl group is positioned between the central carbon carbon at a separation of 209 pm.

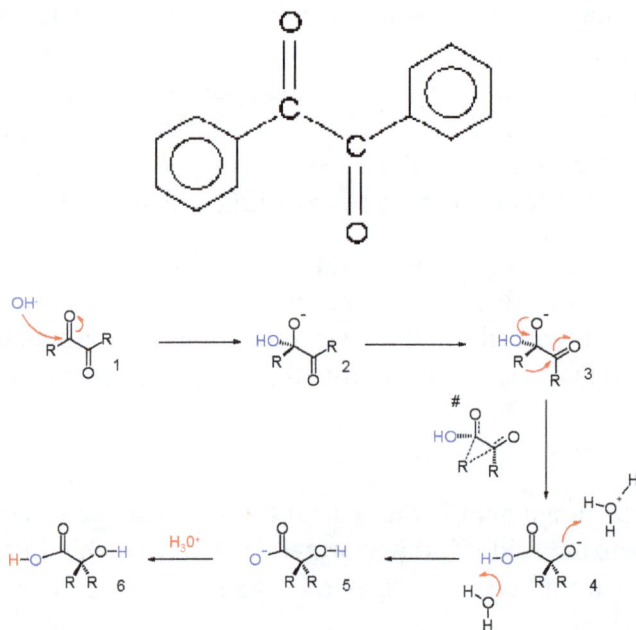

The carboxylic acid in intermediate 4 is less basic than the hydroxyl anion and therefore proton transfer takes place to intermediate 5 which can be protonated in acidic workup to the final α-hydroxy–carboxylic acid 6. Calculations show that an accurate description of the reaction sequence is possible with the participation of 4 water molecules taking responsibility for the stabilization of charge buildup. They also provide a shuttle for the efficient transfer of one proton in the formation of intermediate 5.

Variations

A variation of this reaction occurs in certain steroids. In the so-called D-Homo Rearrangement of Steroids a cyclopentane ring expands to a cyclohexane ring with added base.

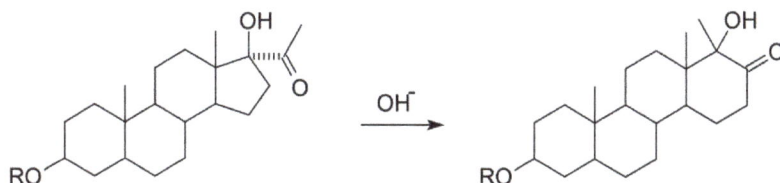

Arndt–Eistert reaction

The Arndt–Eistert synthesis is a series of chemical reactions designed to convert a carboxylic acid to a higher carboxylic acid homologue (i.e. contains one additional carbon atom) and is considered a homologation process. Named for the German chemists Fritz Arndt (1885–1969) and Bernd Eistert (1902–1978), Arndt–Eistert synthesis is a popular method of producing β-amino acids from α-amino acids. Acid chlorides react with diazomethane to give diazoketones. In the presence of a nucleophile (water) and a metal catalyst (Ag_2O), diazoketones will form the desired acid homologue.

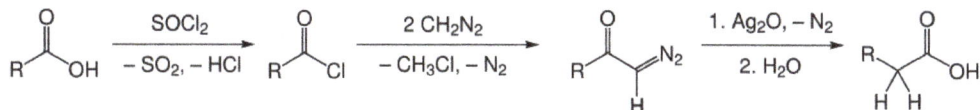

While the classic Arndt–Eistert synthesis uses thionyl chloride to convert the starting acid to an acid chloride, any procedure can be used that will generate an acid chloride.

Diazoketones are typically generated as described here, but other methods such as diazo-group transfer can also apply.

Since diazomethane is toxic and violently explosive, many safer alternatives have been developed, such as the usage of ynolates (Kowalski ester homologation) or diazo(trimethylsilyl)methane.

Reaction Mechanism

The key step in the Arndt–Eistert synthesis is the metal-catalyzed Wolff rearrangement of the diazoketone to form a ketene. In the insertion homologation of t-BOC protected (S)-phenylalanine (2-amino-3-phenylpropanoic acid), t-BOC protected (S)-3-amino-4-phenylbutanoic acid is formed.

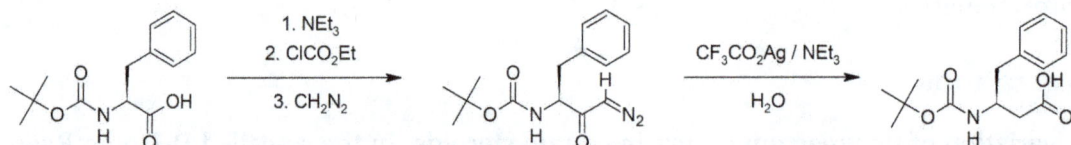

Wolff rearrangement of the α-diazoketone intermediate forms a ketene via a 1,2-rearrangement, which is subsequently hydrolysed to form the carboxylic acid. The consequence of the 1,2-rearrangement is that the methylene group alpha to the carboxyl group in the product is the methylene group from the diazomethane reagent. It has been demonstrated that the rearrangement preserves the stereochemistry of the chiral centre as the product formed from t-BOC protected (S)-phenylalanine retains the (S) stereochemistry with a reported enantiomeric excess of at least 99%.

Heat, light, platinum, silver, and copper salts will also catalyze the Wolff rearrangement to produce the desired acid homologue.

Variations

In the Newman–Beal modification, addition of triethylamine to the diazomethane solution will avoid the formation of α-chloromethylketone side-products.

Rearrangement to Electron Deficient Nitrogen

Hofmann Rearrangement

The Hofmann rearrangement is the organic reaction of a primary amide to a primary amine with one fewer carbon atom.

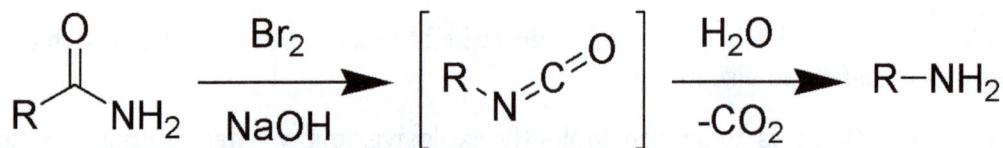

The Hofmann rearrangement.

The reaction is named after its discoverer - August Wilhelm von Hofmann. This reaction is also sometimes called the Hofmann degradation or the Harmon Process, and should not be confused with the Hofmann elimination.

Mechanism

The reaction of bromine with sodium hydroxide forms sodium hypobromite *in situ*, which transforms the primary amide into an intermediate isocyanate. The formation of an intermediate nitrene is not possible because it implies also the formation of an hydroxamic acid as a byproduct, which has never been observed. The intermediate isocyanate is hydrolyzed to a primary amine, giving off carbon dioxide.

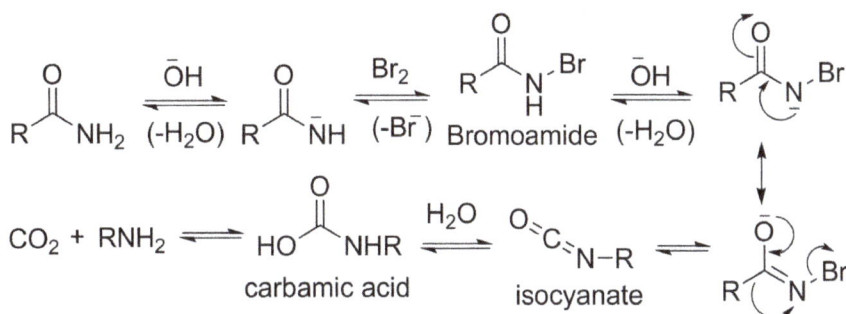

1. Base abstracts an acidic N-H proton, yielding an anion.

2. The anion reacts with bromine in an α-substitution reaction to give an *N*-bromoamide.

3. Base abstraction of the remaining amide proton gives a bromoamide anion.

4. The bromoamide anion rearranges as the R group attached to the carbonyl carbon migrates to nitrogen at the same time the bromide ion leaves, giving an isocyanate.

5. The isocyanate adds water in a nucleophilic addition step to yield a carbamic acid (aka urethane).

6. The carbamic acid spontaneously loses CO_2, yielding the amine product.

Variations

Several reagents can substitute for bromine. Sodium hypochlorite, Lead tetraacetate, *N*-bromosuccinimide, (bis(trifluoroacetoxy)iodo)benzene, and 1,8-diazabicyclo[5.4.0]undec-7-ene (DBU) can affect a Hofmann rearrangement. In the following example, the intermediate isocyanate is trapped by methanol, forming a carbamate.

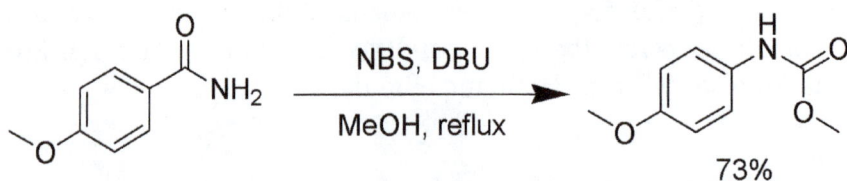

The Hofmann rearrangement using NBS.

In a similar fashion, the intermediate isocyanate can be trapped by *tert*-butyl alcohol, yielding the *tert*-butoxycarbonyl (Boc)-protected amine.

The Hofmann Rearrangement also can be used to yield carbamates from α,β-unsaturated or α-hydroxy amides or nitriles from α,β-Acetylenic amides in good yields (≈70%).

For Amiloride, hypobromous acid was used to effect Hofmann rearrangement.

Applications

- Aliphatic & Aromatic amides are converted into aliphatic and aromatic amines, respectively

- In the preparations of anthranilic acid from phthalimide

- Nicotinic acid is converted into 3-Amino pyridine

- The Symmetrical structure of α-phenyl propanamide does not change after Hofmann reaction.

- Gabapentin from mono-amidation 1,1-cyclohexane diacetic acid anhydride with ammonia to 1,1-cyclohexane diacetic acid mono-amide; followed by 'Hoffmann' rearrangement: U.S. Patent 20,080,103,334

Schmidt Reaction

The Schmidt reaction is an organic reaction in which an azide reacts with a carbonyl group to give an amine or amide, with expulsion of nitrogen. It is named after Karl Friedrich Schmidt (1887–1971), who first reported it in 1924 by successfully converting benzophenone and hydrazoic acid to benzanilide. Surprisingly, the intramolecular reaction wasn't reported until 1991 but has become important in the synthesis of natural products

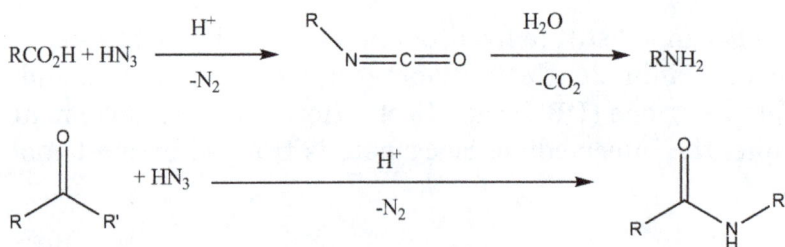

Reaction Mechanism

The carboxylic acid Schmidt reaction starts with acylium ion 1 obtained from protonation and loss of water. Reaction with hydrazoic acid forms the protonated azido ketone 2, which goes through a rearrangement reaction with the alkyl group R, migrating over the C-N bond with expulsion of nitrogen. The protonated isocyanate is attacked by water forming carbamate 4, which after deprotonation loses carbon dioxide to the amine.

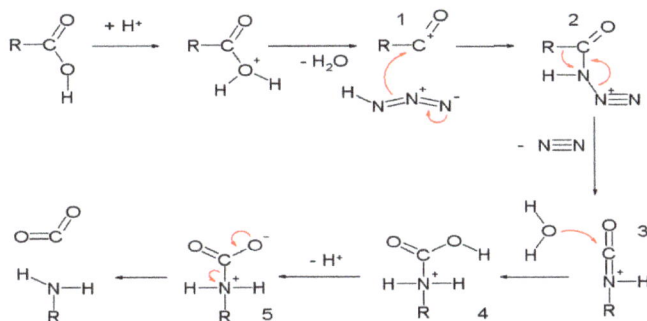

Schmidt reaction mechanism amine formation

The reaction is related to the Curtius rearrangement except that in this reaction the azide is protonated and hence with different intermediates.

In the reaction mechanism for the ketone Schmidt reaction, the carbonyl group is activated by protonation for nucleophilic addition by the azide, forming intermediate 3, which loses water in an elimination reaction to temporary imine 4, over which one of the alkyl groups migrates from carbon to nitrogen with loss of nitrogen. A similar migration is found in the Beckmann rearrangement. Attack by water and proton loss converts 5 to 7, which is a tautomer of the final amide.

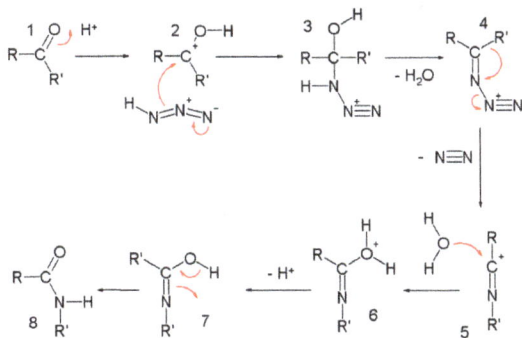

Schmidt reaction mechanism amide formation

Reactions Involving Alkyl Azides

The scope of this reaction has been extended to reactions of carbonyls with alkyl azides $R-N_3$. This extension was first reported by J.H. Boyer in 1955 (hence the name Boyer reaction), for example, the reaction of m-nitrobenzaldehyde with β-azido-ethanol:

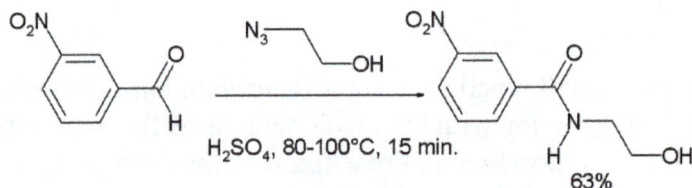

The Boyer reaction

Variations involving intramolecular Schmidt reactions have been known since 1991. These are annulation reactions and have some utility in the synthesis of natural products; such as lactams and alkaloids.

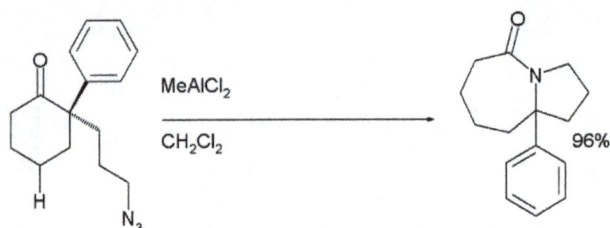

Intramolecular Schmidt Reaction

Lossen Rearrangement

The Lossen rearrangement is the conversion of a hydroxamic acid (1) to an isocyanate (3) via the formation of an O-acyl, sulfonyl, or phosphoryl intermediate hydroxamic acid O-derivative (2) and then conversion to its conjugate base. Here, 4-toluenesulfonyl chloride is used to form a sulfonyl Ortho-derivative of hydroxamic acid.

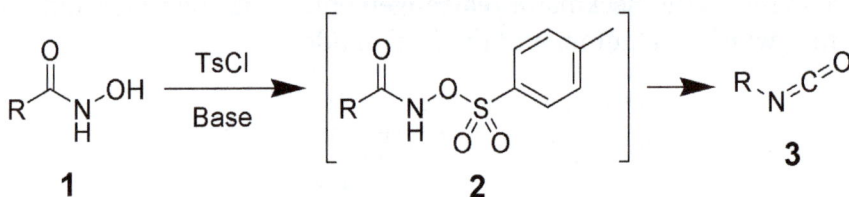

The isocyanate can be used further to generate ureas in the presence of amines (4) or generate amines in the presence of H_2O (5).

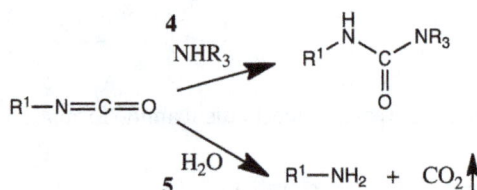

Reaction Mechanism for Lossen Rearrangement

The mechanism below begins with an O-acylated hydroxamic acid derivative that is treated with base to form an isocyanate that generates an amine and CO_2 gas in the presence of H_2O. The hydroxamic acid acid derivative is first converted to its conjugate base by abstraction of a hydrogen by a base. Spontaneous rearrangement kicks off a carboxylate anion to produce the isocyanate intermediate. The isocyanate in the presence H_2O hydrolyzes and then decarboxylation via abstraction of a hydrogen by a base generates an amine and CO_2 gas.

Overall:

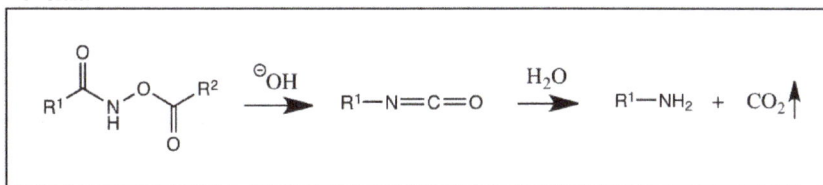

Mechanism:

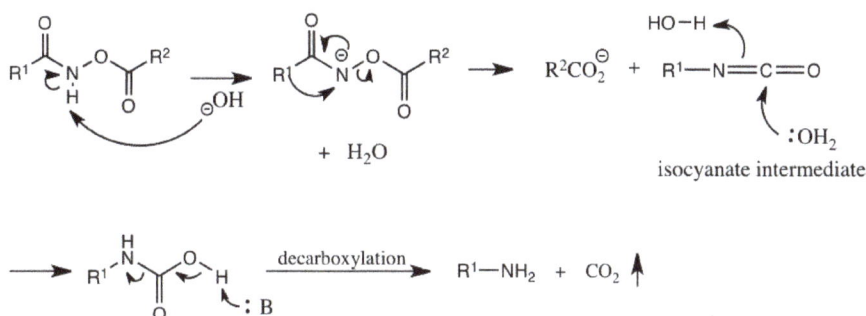

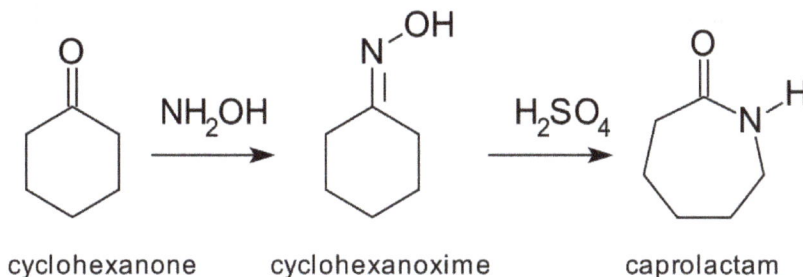

Hydroxamic acids are commonly synthesized from their corresponding esters.

Beckmann Rearrangement

The Beckmann rearrangement, named after the German chemist Ernst Otto Beckmann (1853–1923), is an acid-catalyzed rearrangement of an oxime to an amide. Cyclic oximes yield lactams.

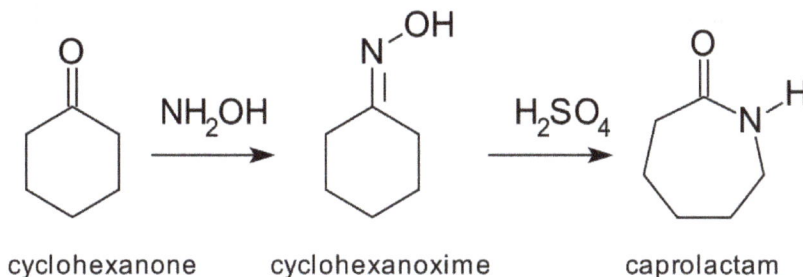

This example reaction starting with cyclohexanone, forming the reaction intermediate cyclohexanone oxime and resulting in caprolactam is one of the most important applications of the Beckmann rearrangement, as caprolactam is the feedstock in the production of Nylon 6.

The Beckmann solution consists of acetic acid, hydrochloric acid and acetic anhydride, and was widely used to catalyze the rearrangement. Other acids, such as sulfuric acid or polyphosphoric acid, can also be used. Sulfuric acid is the most commonly used acid for commercial lactam production due to its formation of an ammonium sulfate by-product when neutralized with ammonia. Ammonium sulfate is a common agricultural fertilizer providing nitrogen and sulfur.

Reaction Mechanism

The reaction mechanism of the Beckmann rearrangement is in general believed to consist of an alkyl migration with expulsion of the hydroxyl group to form a nitrilium ion followed by hydrolysis:

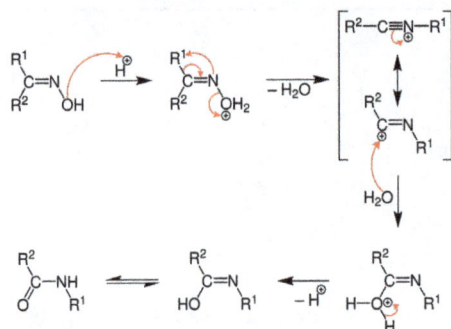

In one study, the mechanism is established in silico taking into account the presence of solvent molecules and substituents. The rearrangement of acetone oxime in the Beckmann solution involves three acetic acid molecules and one proton (present as an oxonium ion). In the transition state leading to the iminium ion (σ-complex), the methyl group migrates to the nitrogen atom in a concerted reaction and the hydroxyl group is expulsed. The oxygen atom in the hydroxyl group is stabilized by the three acetic acid molecules. In the next step the electrophilic carbon atom in the nitrilium ion is attacked by water and the proton is donated back to acetic acid. In the transition state leading to the N-methyl acetimidic acid, the water oxygen atom is coordinated to 4 other atoms. In the third step, an isomerization step protonates the nitrogen atom leading to the amide.

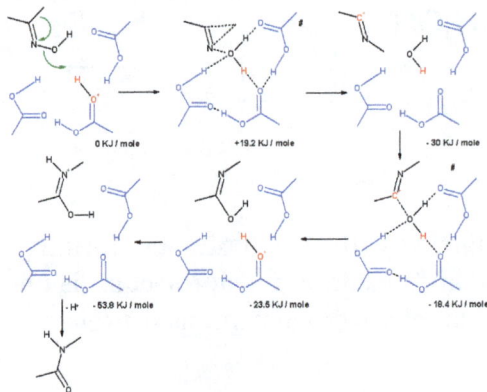

The same computation with a hydroxonium ion and 6 molecules of water has the same result, but, when the migrating substituent is phenyl in the reaction of acetophenone oxime with protonated acetic acid, the mechanism favors the formation of an intermediate three-membered π-complex. This π-complex is again not found in the $H_3O^+(H_2O)_6$.

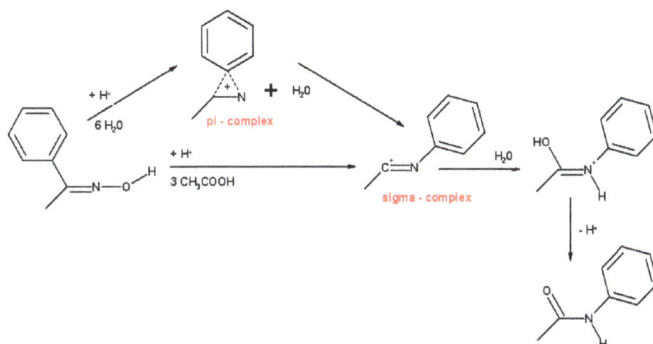

With the cyclohexanone-oxime, the relief of ring strain results in a third reaction mechanism, leading directly to the protonated caprolactam in a single concerted step without the intermediate formation of a π-complex or σ-complex.

Cyanuric Chloride Assisted Beckmann Reaction

Beckmann reaction is known to be catalyzed by cyanuric chloride and zinc chloride co-catalyst. For example, cyclododecanone can be converted to the corresponding lactam, a monomer for the production of Nylon 12.

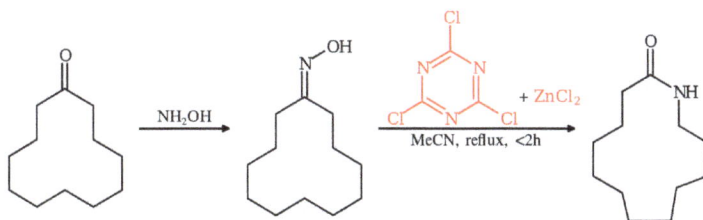

The reaction mechanism for this reaction is based on a catalytic cycle with cyanuric chloride activating the hydroxyl group via a nucleophilic aromatic substitution. The reaction product is dislodged and replaced by new reactant via an intermediate Meisenheimer complex.

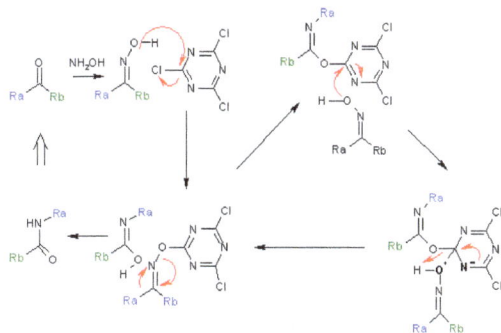

Beckmann Fragmentation

When the oxime has a quaternary carbon atom in an anti position to the hydroxyl group a fragmentation occurs forming a nitrile:

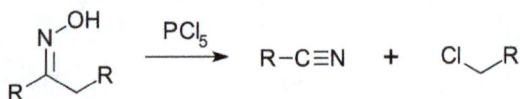

The fluorine donor in this fragmentation reaction is diethylaminosulfur trifluoride (DAST):

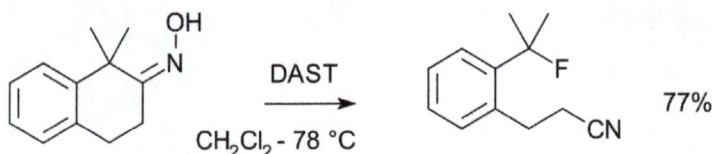

Semmler–Wolff Reaction

The oxime of cyclohexenone with acid forms aniline in a dehydration – aromatization reaction called the Semmler–Wolff reaction or Wolff aromatization

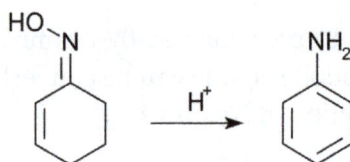

Applications in Drug Synthesis

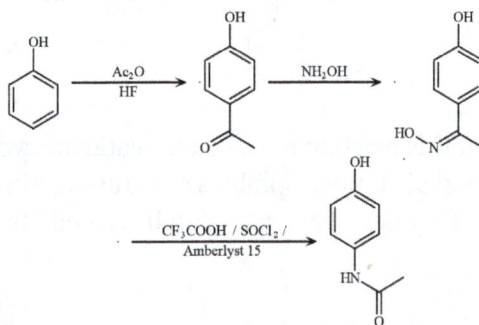

This route also involves the Beckmann Rearrangement, like the conversion from cyclohexanone to caprolactam.

An industrial synthesis of paracetamol developed by Hoechst–Celanese involves the conversion of ketone to a ketoxime with hydroxylamine.

Other known examples of the B.R. include:

- DHEA, Benazepril, Ceforanide precursor, Elanzepine, as well as for 17-azapro-gesterone.

B.R. on Anthraquinone is used to make Elantrine, Prazepine, Enprazepine. Etazepine.

Hofmann Rearrangement

This rearrangement provides an effective method for the synthesis of primary aliphatic and aromatic amines from primary amides.

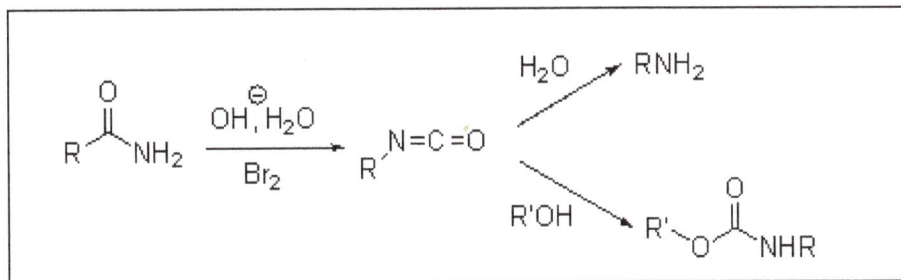

Mechanism

Treatment of amide with sodium hypobromite gives N -bromo-amide which reacts with base to afford a conjugate base within which rearrangement takes place to give iso-cyanate. The formed isocyanate may be isolated in anhydrous conditions or it can be converted into amine by aqueous workup.

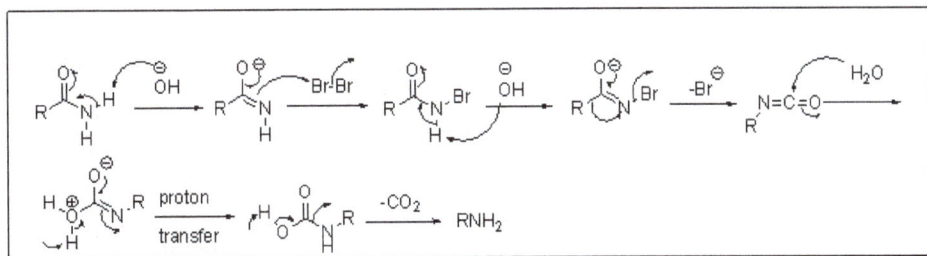

The workup can also be with alcohol or amine to give urethane or urea, respectively.

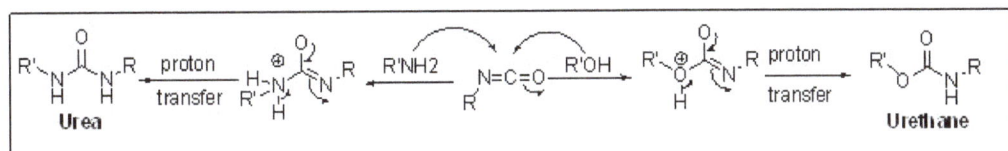

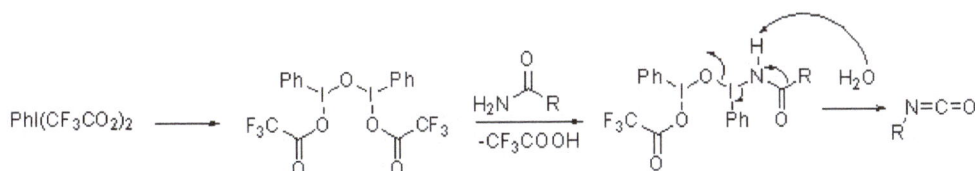

Examples

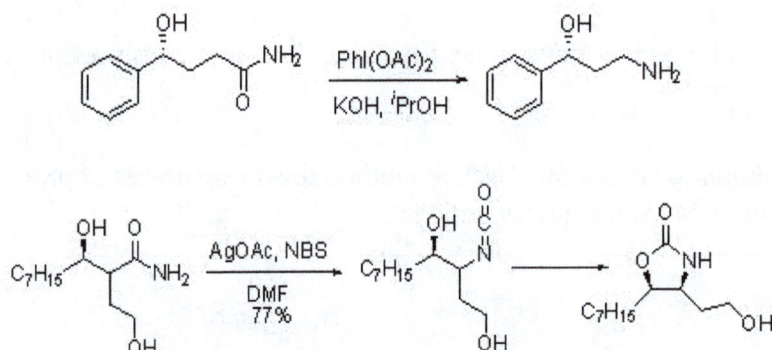

Curtius Rearrangement

This rearrangement describes the transformation of acyl azide into isocyanate by decomposition on heating and its application for the synthesis of primary amines, urethanes and ureas as presented in Hofmann rearrangement.

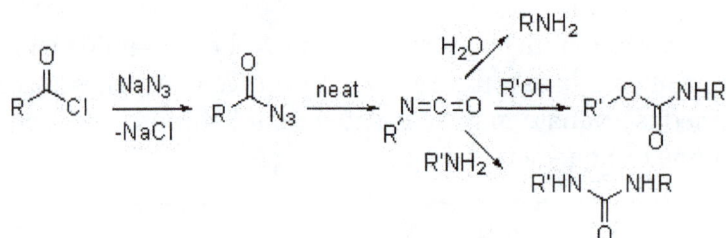

Mechanism

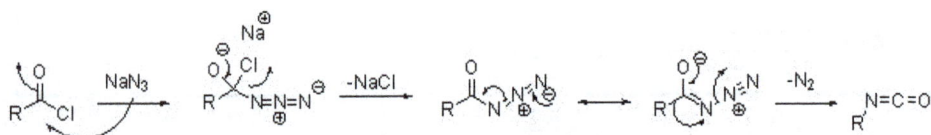

Examples

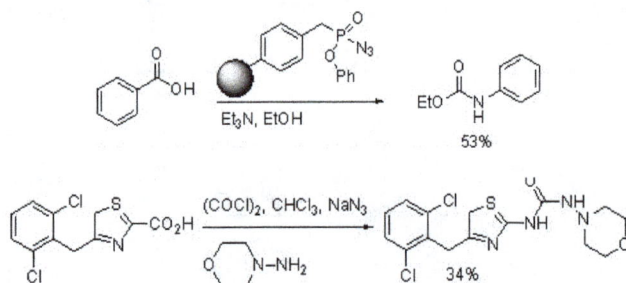

Schmidt Rearrangement

Carboxylic acid reacts with hydrazoic acid in the presence of conc. H_2SO_4 to give acid azide which is present in the form of conjugate acid eliminates nitrogen to afford isocyanate that could be converted into amine as reported Hofmann rearrangement.

The reaction is also effective with aldehydes, ketones, tertiary alcohols and substituted alkenes.

Mechanism

Examples

Lossen Rearrangement

Ester of hydroxamic acid reacts with base to give isocyanate that could be converted into amine as shown in Hofmann rearrangement.

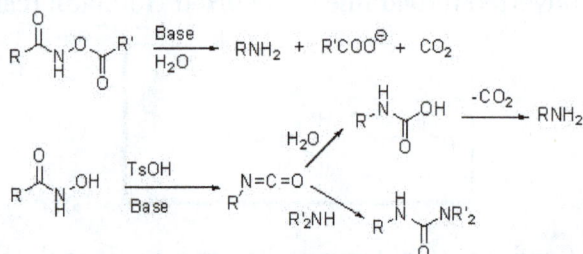

Mechanism

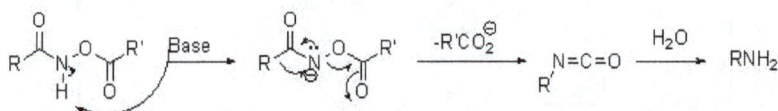

Examples

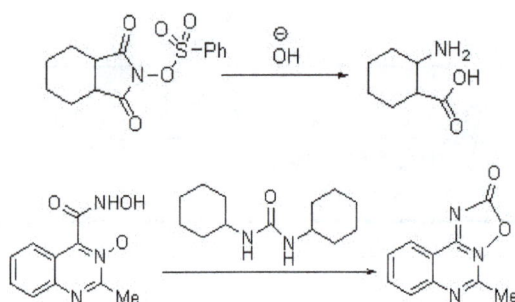

All these four rearrangements have common intermediate isocyanate forming from different substrate precursors..

Beckmann Rearrangement

Oximes rearranges in acidic conditions to give amides. The reaction is intramolecular and stereospecific: the substituent trans to the leaving groups migrates.

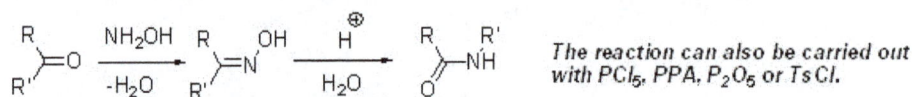

The reaction can also be carried out with PCl$_5$, PPA, P$_2$O$_5$ or TsCl.

An interesting application of this method is the synthesis caprolactam from cyclohexanone oxime. Caprolactam is the substrate precursor for nylon preparation.

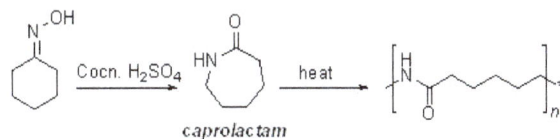

caprolactam

Mechanism

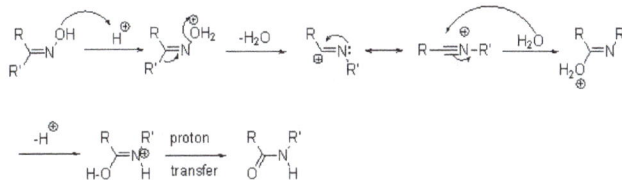

Examples

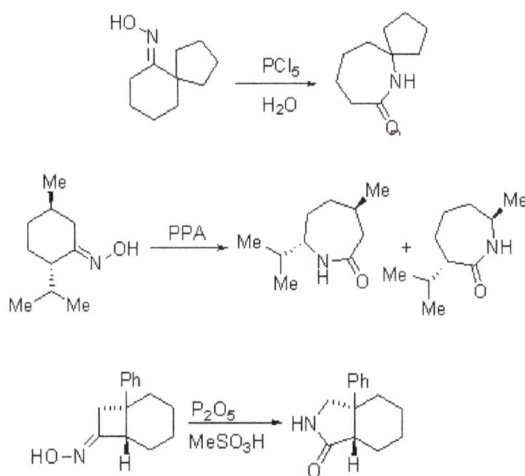

The rearrangement of amidoximes lead to the formation of urea derivatives which is called the Tiemann Rearrangement

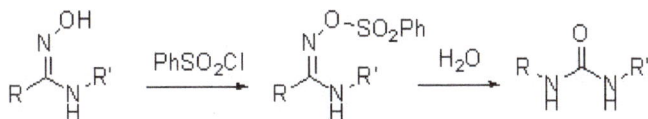

Rearrangement to Electron Deficient Oxygen

Baeyer Villiger Reaction

Treatment of ketones with peroxyacid gives ester. The reaction is effective with acid or base and the mechanism is closely related to pinacol rearrangement: nucleophilic attack by the peroxyacid on the carbonyl group gives an intermediate that rearranges with the expulsion of the anion of the acid.

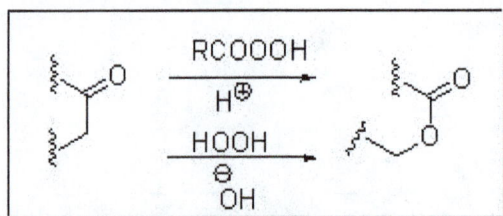

Mechanisms

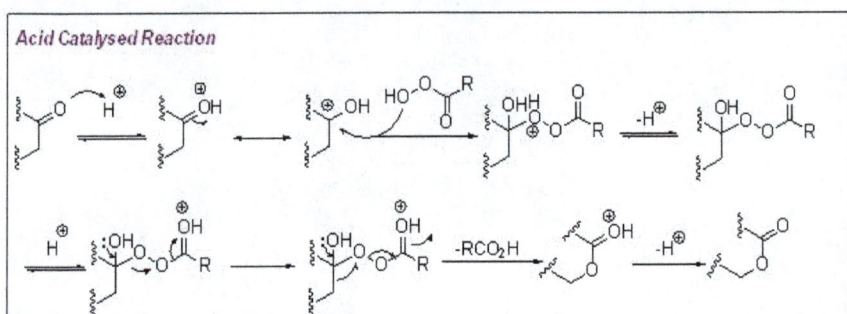

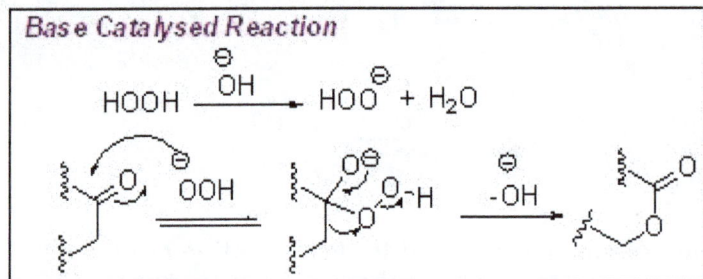

Migratory Aptitude: 3° 2°> PhCH₂> Ph > 1°> Me > H.

Examples

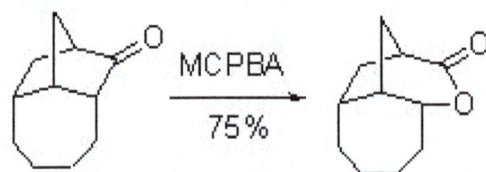

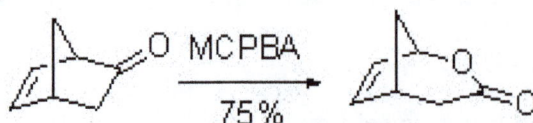

Hydroperoxide Rearrangement

Tertiary hydroperoxide with acid undergoes rearrangement to give ketone and alcohol or phenol. The mechanism is similar to that of Baeyer-Villiger reaction. For example, cumene forms hydroperoxide by autoxidation which is with acid rearranges to give phenol and acetone.

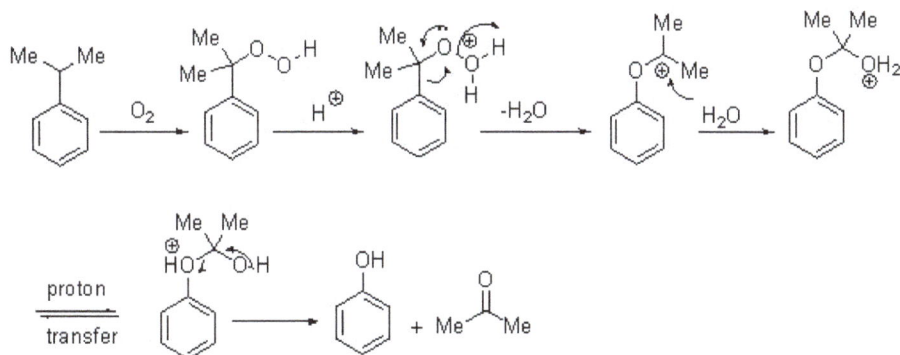

Dakin Reaction

Benzaldehyde or acetophenone bearing hydroxyl substituent in the ortho or para position proceed rearrangement to give catechol or quinol , respectively. The reaction is performed in the presence of alkaline hydrogen peroxide and the mechanism is similar to that of Baeyer-Villiger reaction.

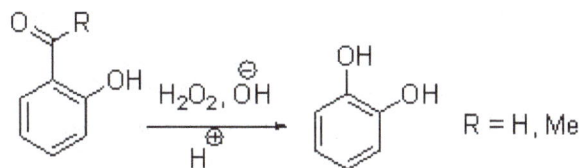

Proposed Mechanism

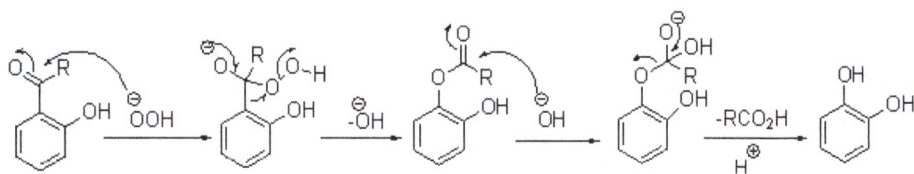

Examples

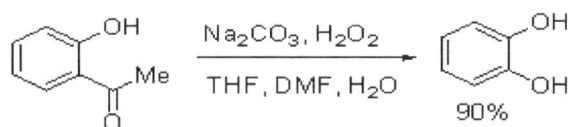

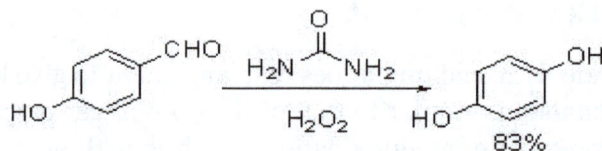

Rearrangement to Electron-Rich Carbon

This group of reaction has been less explored, and is less of synthetic importance compared to the rearrangements to electron deficient carbons. The rearrangements to electron deficient hetero atom may be generally explained as:

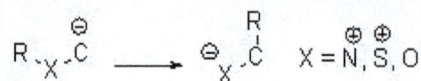

Stevens Rearrangement

Quaternary ammonium salt which has β -hydrogen proceeds E_2 (Hofmann) elimination with base.

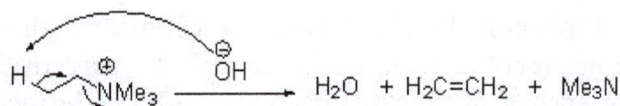

In case of quaternary ammonium salts containing β -ketone or ester or aryl group, an a -hydrogen is removed by base to give an ylide and then the rearrangement occurs.

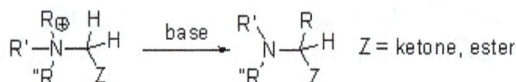

Mechanism

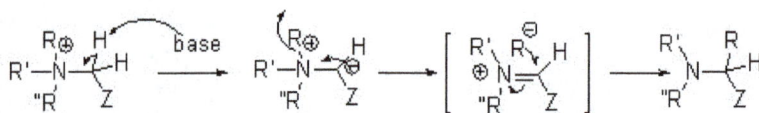

Examples

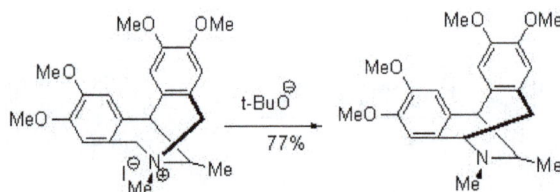

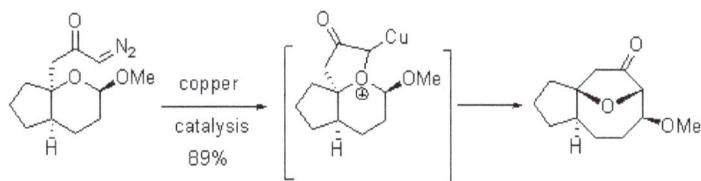

Sommelet-Hauser Rearrangement

In the absence of β -carbonyl group, the α -hydrogen is too weakly acidic for hydroxide ion induced rearrangement. Thus, a strong base, such as amide ion in liquid ammonia, is to be used, when the rearrangement takes a different course: instead of [1,2] shift (Steven's rearrangement), a [3,2]-sigmatropic rearrangement takes place which is called Sommelet- Hauser rearrangement.

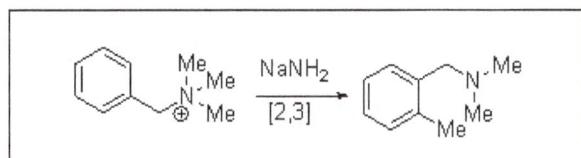

Mechanism

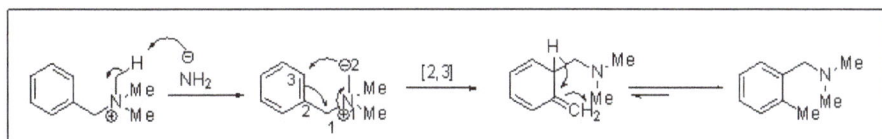

There can be competition between Stevens and Sommelet-Hauser rearrangement mechanisms.

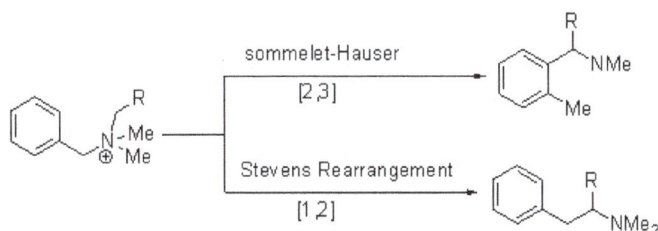

Examples

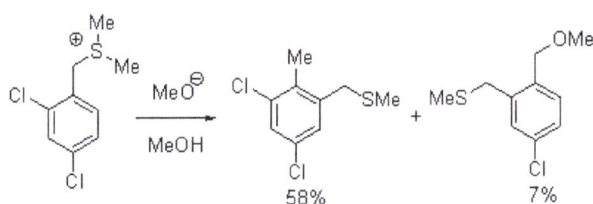

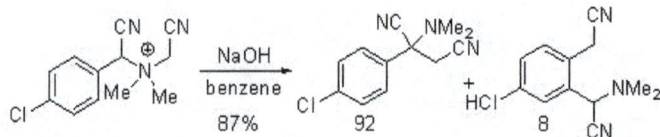

Wittig Rearrangement

Ethers undergo [1,2]-sigmatropic rearrangement in the presence of strong base such as amide ion or phenyllithium to give more stable oxyanion. The mechanism is analogous to that of Stevens rearrangement.

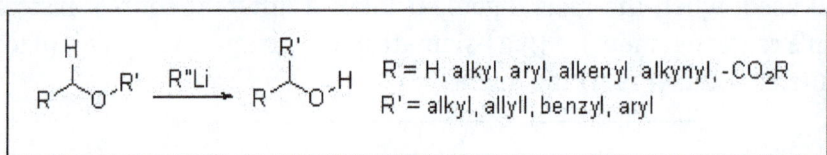

R = H, alkyl, aryl, alkenyl, alkynyl, -CO₂R
R' = alkyl, allyll, benzyl, aryl

Mechanism

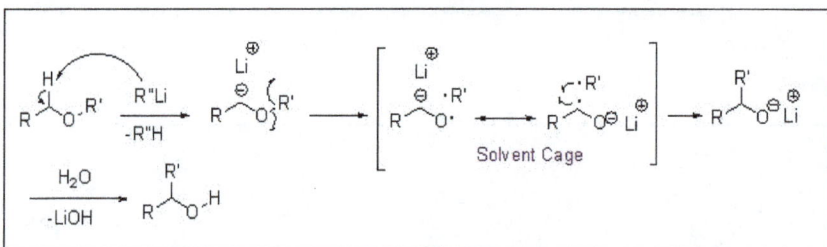

Examples

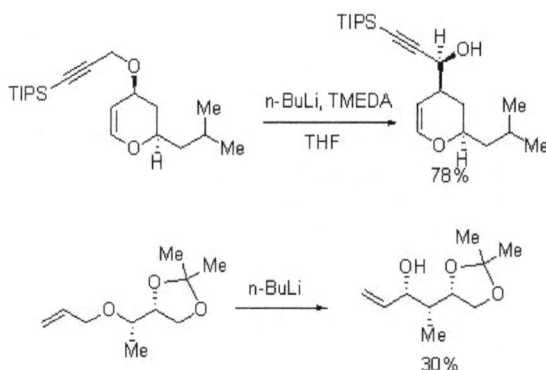

Favorskii Rearrangement

α -Haloketones with base afford enolates which rearrange to give esters via cyclopropanones.

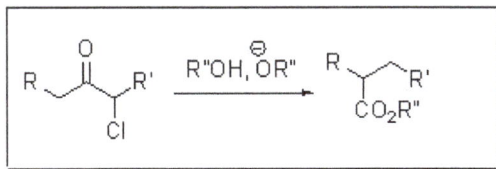

Mechanism

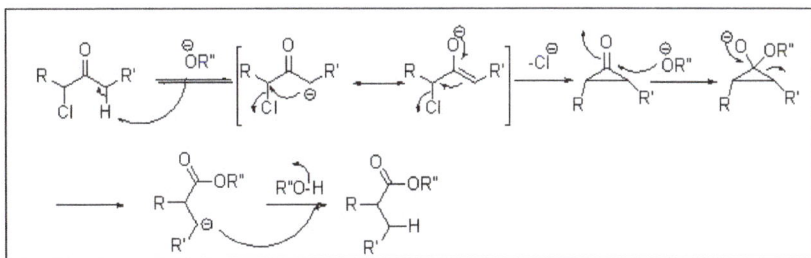

The direction of ring opening of cyclopropanone is determined by the more stable carbanion, formed in the reaction.

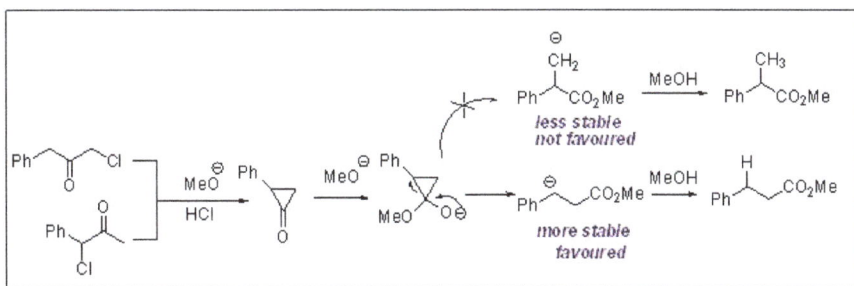

Examples

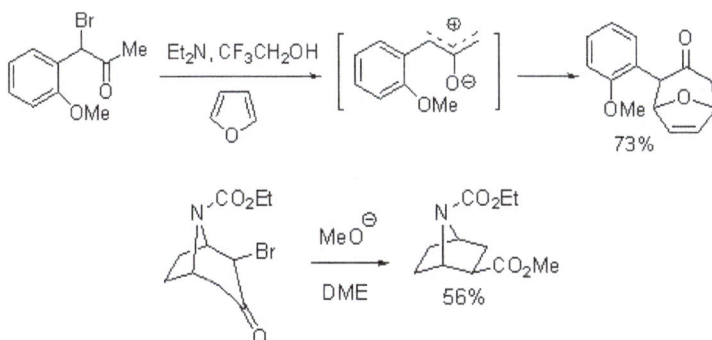

Stevens Rearrangement

The Stevens rearrangement in organic chemistry is an organic reaction converting quaternary ammonium salts and sulfonium salts to the corresponding amines or sulfides in presence of a strong base in a 1,2-rearrangement.

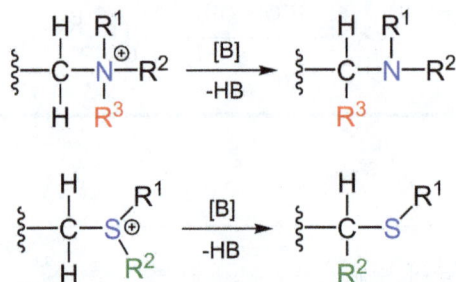

The reactants can be obtained by alkylation of the corresponding amines and sulfides. The substituent R next the amine methylene bridge is an electron-withdrawing group.

The original 1928 publication by Thomas S. Stevens concerned the reaction of *1-phenyl-2- (N, N-dimethyl) ethanone* with benzyl bromide to the ammonium salt followed by the rearrangement reaction with sodium hydroxide in water to the rearranged amine.

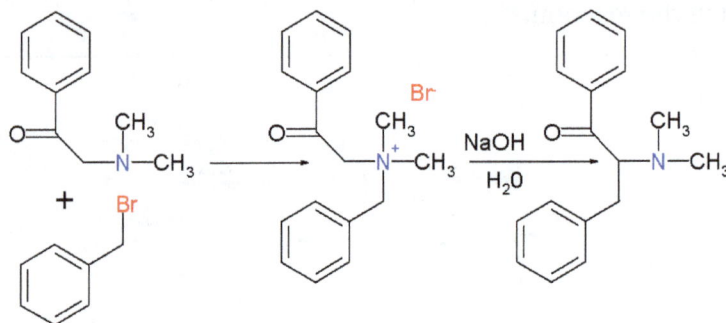

A 1932 publication described the corresponding sulfur reaction.

Reaction Mechanism

The reaction mechanism of the Stevens rearrangement is one of the most controversial reaction mechanism in organic chemistry. Key in the reaction mechanism for the Stevens rearrangement (explained for the nitrogen reaction) is the formation of an ylide after deprotonation of the ammonium salt by a strong base. Deprotonation is aided by electron-withdrawing properties of substituent R. Several reaction modes exist for the actual rearrangement reaction.

A concerted reaction requires an antarafacial reaction mode but since the migrating group displays retention of configuration this mechanism is unlikely.

In an alternative reaction mechanism the N–C bond of the leaving group is homolytically cleaved to form a di-radical pair (3a). In order to explain the observed retention of configuration, the presence of a solvent cage is invoked. Another possibility is the formation of a cation-anion pair (3b), also in a solvent cage. Recently the elimination recombination coupling mechanism opens a new approach to understand the formation of normal and abnormal product in the stevens rearrangement

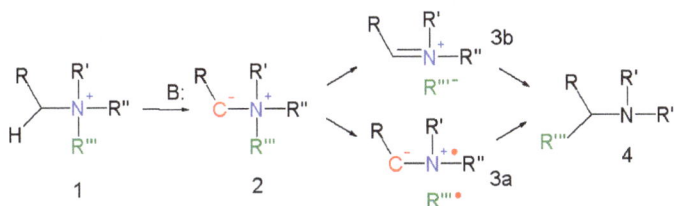

Scope

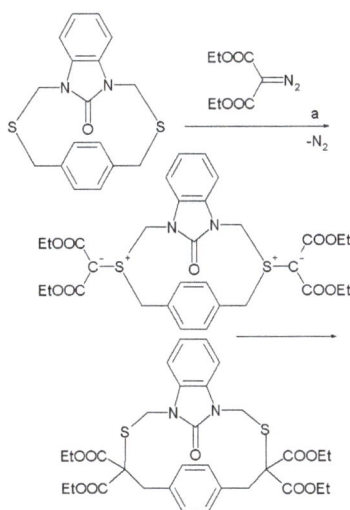

Competing reactions are the Sommelet-Hauser rearrangement and the Hofmann elimination.

In one application a double-Stevens rearrangement expands a cyclophane ring. The ylide is prepared in situ by reaction of the diazo compound *ethyl diazomalonate* with a sulfide catalyzed by dirhodium tetraacetate in refluxing xylene.

Enzymatic Reaction

Recently, γ-butyrobetaine hydroxylase, an enzyme that is involved in the human carnitine biosynthesis pathway, was found to catalyze a C-C bond formation reaction in a fashion analogous to a Stevens' type rearrangement. The substrate for the reaction is meldonium.

Sommelet–Hauser Rearrangement

The Sommelet–Hauser rearrangement (named after M. Sommelet and Charles R.

Hauser) is a rearrangement reaction of certain benzyl quaternary ammonium salts. The reagent is sodium amide or another alkali metal amide and the reaction product a *N,N*-dialkylbenzylamine with a new alkyl group in the aromatic ortho position. For example, benzyltrimethylammonium iodide, [(C$_6$H$_5$CH$_2$)N(CH$_3$)$_3$]I, rearranges in the presence of sodium amide to yield the *o*-methyl derivative of *N,N*-dimethylbenzylamine.

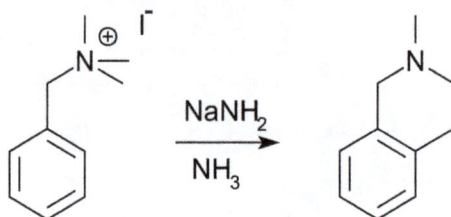

Mechanism

The benzylic methylene proton is acidic and deprotonation takes place to produce the benzylic ylide (1). This ylide is in equilibrium with a second ylide that is formed by deprotonation of one of the ammonium methyl groups (2). Though the second ylide is present in much smaller amounts, it undergoes a 2,3-sigmatropic rearrangement and subsequent aromatization to form the final product (3).

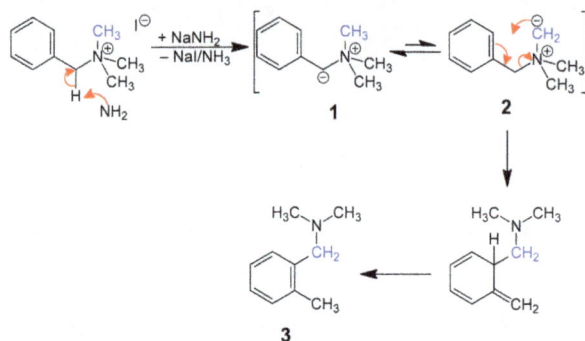

The Stevens rearrangement is a competing reaction.

1,2-Wittig Rearrangement

A 1,2-Wittig rearrangement is a categorization of chemical reactions in organic chemistry, and consists of a 1,2-rearrangement of an ether with an alkyllithium compound. The reaction is named for Nobel Prize winning chemist Georg Wittig.

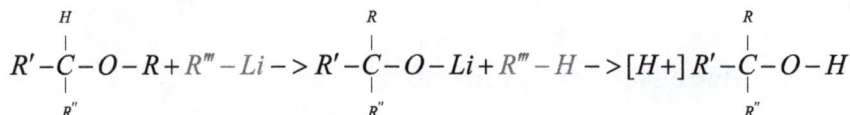

$$R' - \overset{\overset{\displaystyle H}{|}}{\underset{\underset{\displaystyle R''}{|}}{C}} - O - R + R''' - Li -> R' - \overset{\overset{\displaystyle R}{|}}{\underset{\underset{\displaystyle R''}{|}}{C}} - O - Li + R''' - H -> [H+] \, R' - \overset{\overset{\displaystyle R}{|}}{\underset{\underset{\displaystyle R''}{|}}{C}} - O - H$$

The intermediate product is an alkoxy lithium salt and the final product an alcohol. When R" is a good leaving group and electron withdrawing functional group such as a cyanide (CN) group, this group is eliminated and the corresponding ketone is formed.

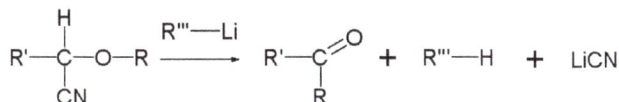

$$R'-\underset{\underset{CN}{|}}{\overset{\overset{H}{|}}{C}}-O-R \xrightarrow{R'''-Li} R'-\underset{\underset{R}{|}}{C}\overset{\nearrow O} + R'''-H + LiCN$$

Reaction Mechanism

The reaction mechanism centers on the formation of a free radical pair with lithium migrating from the carbon atom to the oxygen atom. The R radical then recombines with the ketyl.

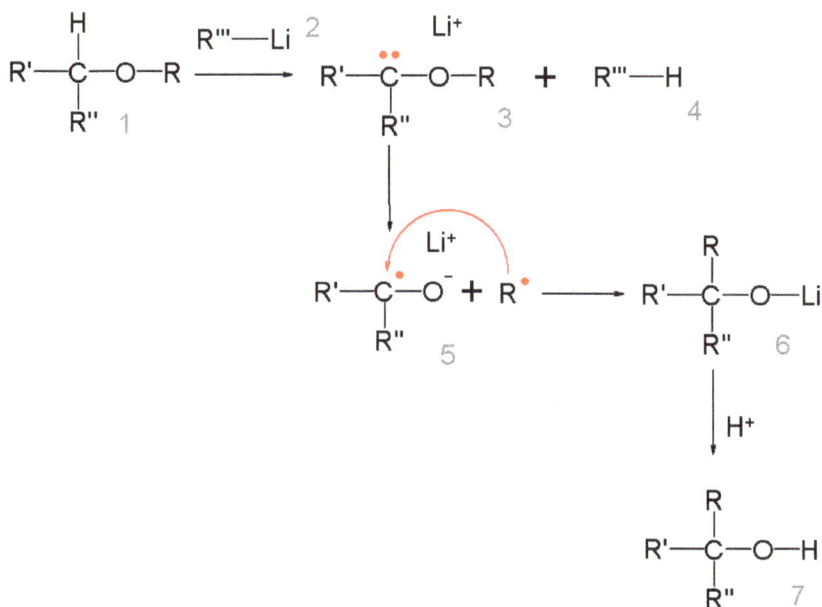

$$R'-\underset{\underset{R''}{|}}{\overset{\overset{H}{|}}{C}}-O-R \xrightarrow{R'''-Li \;\; 2} \overset{Li^+}{R'-\underset{\underset{R''}{|}}{C}-O-R} + R'''-H$$

$$\overset{Li^+}{R'-\underset{\underset{R''}{|}}{\overset{\cdot}{C}}-O^- + R^\cdot} \longrightarrow R'-\underset{\underset{R''}{|}}{\overset{\overset{R}{|}}{C}}-O-Li$$

$$\xrightarrow{H^+} R'-\underset{\underset{R''}{|}}{\overset{\overset{R}{|}}{C}}-O-H$$

The alkyl group migrates in the order of thermodynamical stability methyl < primary alkyl < secondary alkyl < tertiary alkyl in this is line with the radical mechanism. The radical-ketyl pair is short lived and due to a solvent cage effect some isomerizations take place with retention of configuration.

With certain allyl aryl ethers a competing reaction mechanism takes place. The reaction of *allyl phenyl ether* 1 with *sec-butyllithium* at −78 °C gives the lithiated intermediate 2 which on heating to −25 °C only shows the rearranged product 5 but not 4 after trapping the lithium alkoxide with trimethylsilyl chloride. This result rules out a radical-ketyl intermediate 3a in favor of the Meisenheimer complex 3b. Additional evidence for this mechanism is provided by the finding that with a para tert-butyl substituent the reaction is retarded.

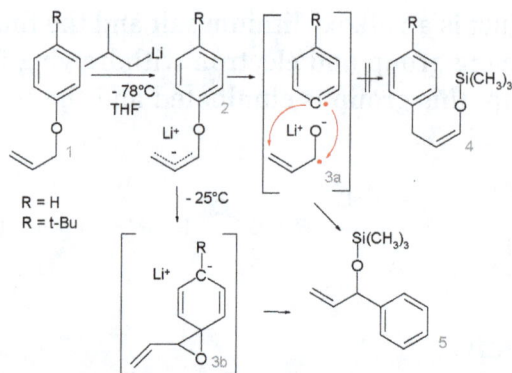

The reaction is a formal dyotropic reaction.

Favorskii Rearrangement

The Favorskii rearrangement, named for the Russian chemist Alexei Yevgrafovich Favorskii, is most principally a rearrangement of cyclopropanones and α-halo ketones which leads to carboxylic acid derivatives. In the case of cyclic α-halo ketones, the Favorskii rearrangement constitutes a ring contraction. This rearrangement takes place in the presence of a base, sometimes hydroxide, to yield a carboxylic acid but most of the time either an alkoxide base or an amine to yield an ester or an amide, respectively. α,α'-Dihaloketones eliminate HX under the reaction conditions to give α,β-unsaturated carbonyl compounds.

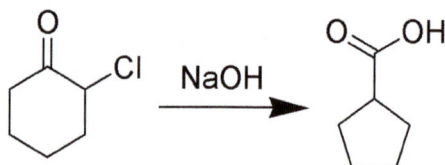

Reaction Mechanism

The reaction mechanism is thought to involve the formation of an enolate on the side of the ketone away from the chlorine atom. This enolate cyclizes to a cyclopropanone intermediate which is then attacked by the hydroxide nucleophile.

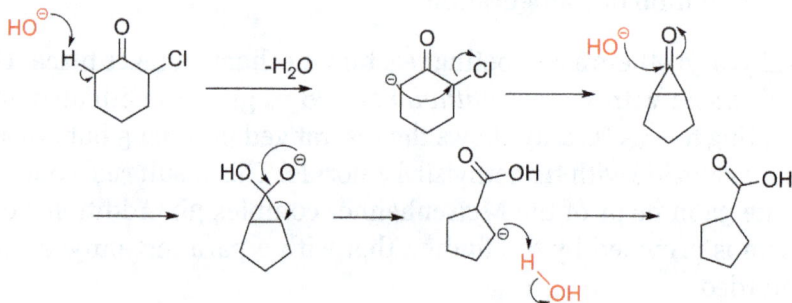

Usage of alkoxide anions such as sodium methoxide, instead of sodium hydroxide, yields the ring-contracted ester product.

Wallach Degradation

In the related Wallach degradation (Otto Wallach, 1918) not one but two halogen atoms flank the ketone resulting in a new contracted ketone after oxidation and decarboxylation

Photo-Favorskii Reaction

The reaction type also exists as a photochemical reaction. The photo-Favorskii reaction has been used in the photochemical unlocking of certain phosphates (for instance those of ATP) protected by so-called *p-hydroxyphenacyl* groups. The deprotection proceeds through a triplet diradical (3) and a dione spiro intermediate (4) although the latter has thus far eluded detection.

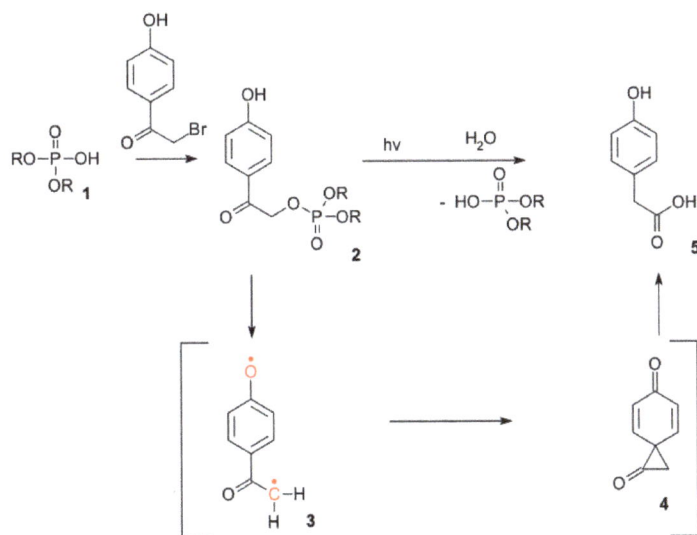

Aromatic Rearrangements

A number of rearrangements occur in aromatic compounds of the type:

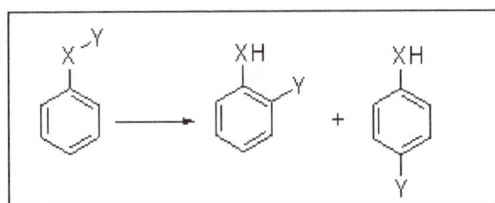

X is usually nitrogen or oxygen. Both intermolecular and intramolecular migrations are known.

Intermolecular Migration from Nitrogen to Carbon

Aniline derivatives readily proceed rearrangement on treatment with acid. First, the formation of conjugate acid of the amine takes place which then eliminates the electrophilic species that reacts at the activated ortho or para position of the aromatic ring.

N-Haloanilides (Orton Rearrangement)

Treatment of N -chloroacetanilide with hydrochloric acid affords a mixture of ortho and para -chloracetanilides in the same proportions as in the direct chlorination of acetanilide.

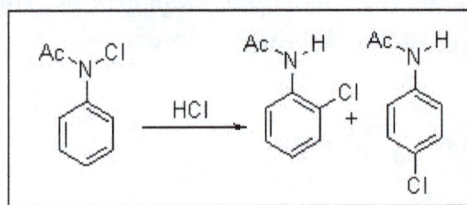

Mechanisms

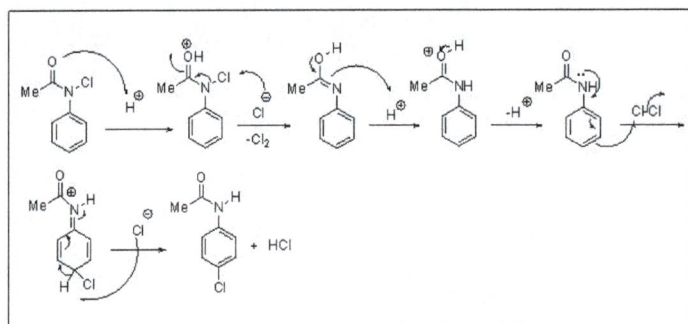

N-Alkyl-N-nitrosoanilines (Fisher-Hepp Rearrangement)

The conjugate acid of the amine releases nitrosonium ion which reacts at para -position to give the p -nitroso product.

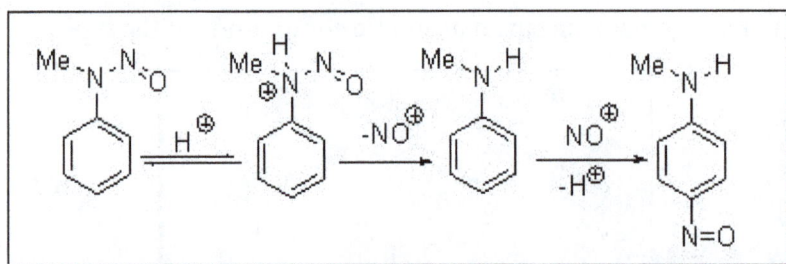

Mechanism

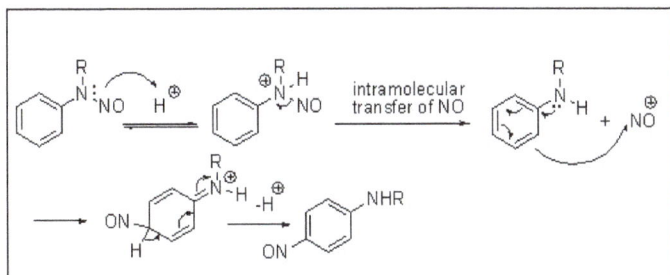

N-Arylazoanilines

N-Arylazoanilines undergo rearrangement in presence of an acid to produce 4-(2-aryldi-azenyl)aniline. On treatment with acid, aryldiazonium ion is formed from the conjugate acid of amine, which migrates to the para position almost selectively.

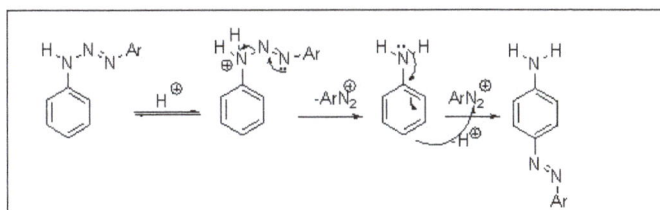

N-Alkylanilines (Hofmann-Martius Rearrangement)

The mechanism of this rearrangement is same as described above, except the require-ment of higher temperature (250-300°C).

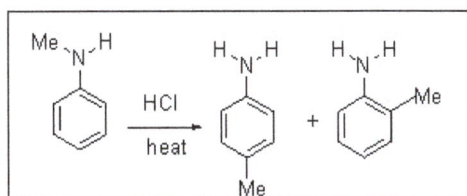

Mechanism

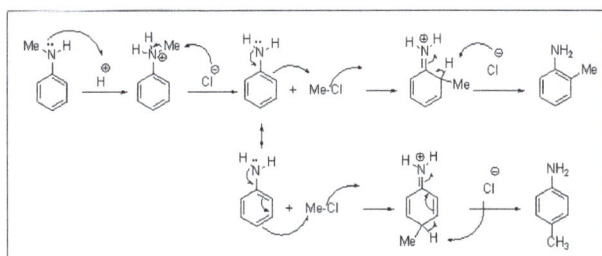

N-Arylhydroxylamines (Bamberger Rearrangement)

Arylhydroxyamines with acid undergoes rearrangement to give aminophenols. Mechanism of this reaction is different from those described above. In this rearrangement, the conjugate acid of the hydroxylamine undergoes nucleophilic attack by the solvent.

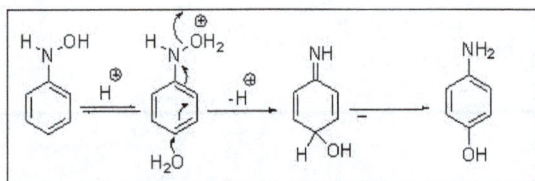

Examples

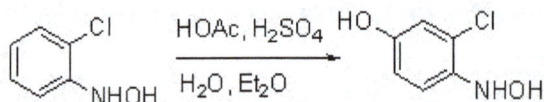

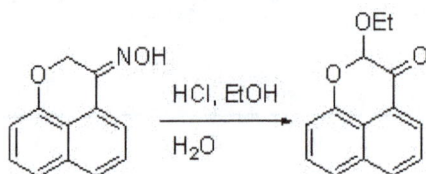

Fries Rearrangement

Aryl esters with Lewis acid undergo rearrangement to give phenols having keto substituent at ortho and para positions. The complex between the ester and Lewis acid gives an acylium ion which reacts at the ortho and para positions as in Friedel-Crafts acylation.

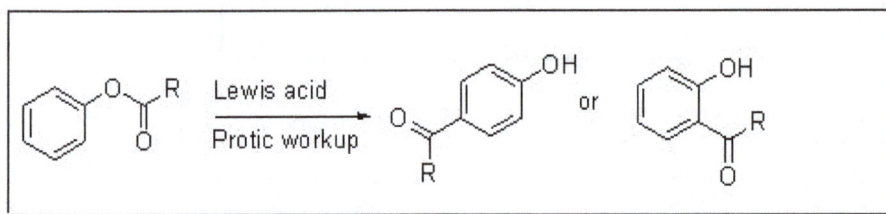

Mechanism

In general, low temperature favors the formation of para- product (kinetic control) and high temperature lead to the formation ortho -product (thermodynamic control).

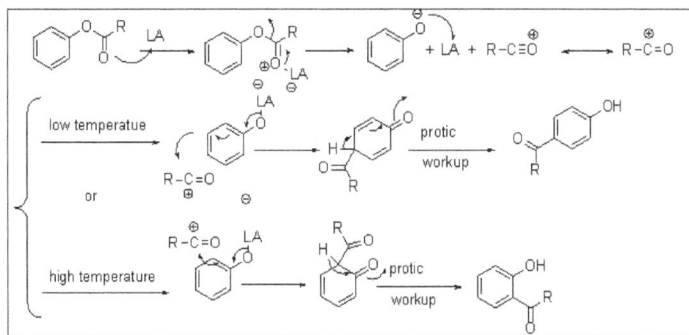

Examples

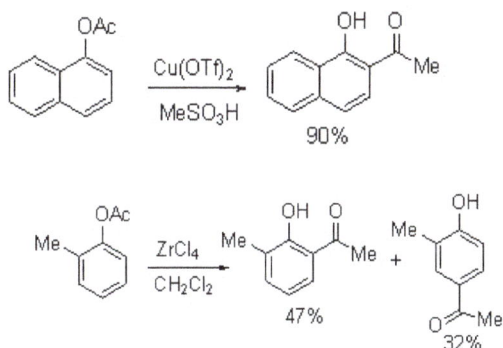

Intramolecular Migration from Nitrogen to Carbon

The mechanisms of these reactions are not fully understood.

Phenylnitramines

These compounds on heating with acid rearrange to give mainly the o -nitro-derivative. For example,

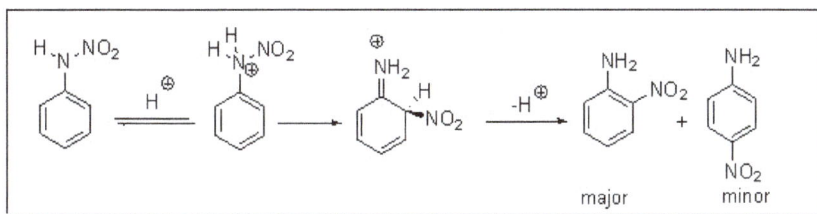

Phenylsulfamic Acids

These compounds rearrange on heating to give o -sulfonic acid derivative that further rearranges at high temperature to afford p -sulfonic acid derivatives. For example,

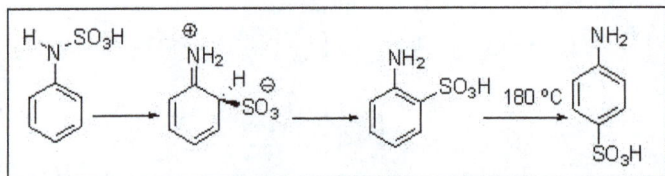

Hydrazobenzenes

These compounds undergo [5,5]-sigmatropic rearrangement in the presence of acid to give benzidines.

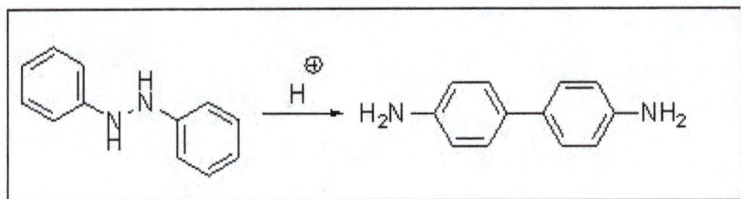

Mechanism

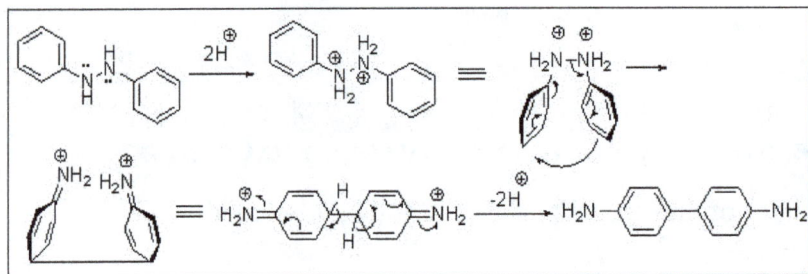

Examples

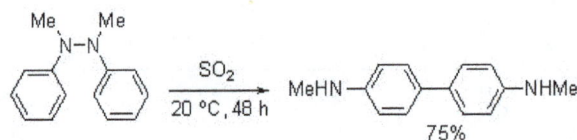

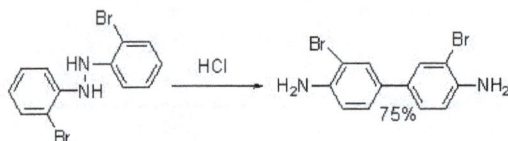

Claisen Rearrangement

Aryl allyl ethers undergo [3,3]-sigmatropic rearrangement on being heated to allylphenols.

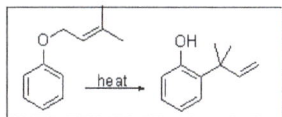

Mechanism

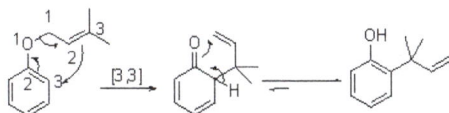

If the ortho position is blocked, rearrangement continues to give para -product.

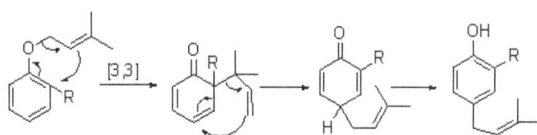

Examples

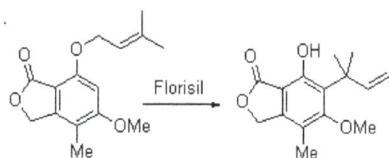

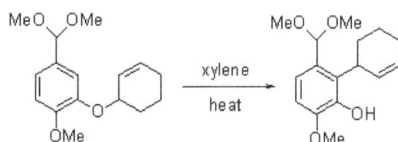

Thermal Rearrangement of Aromatic Hydrocarbons

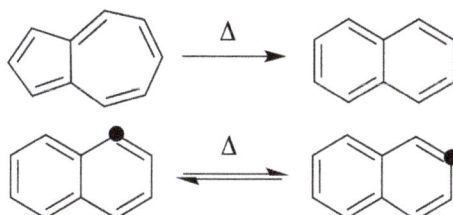

● ^{13}C-label

Thermal isomerization of azulene to naphthalene;
Bottom) Thermal autonomerization of naphthalene.

Thermal rearrangements of aromatic hydrocarbons are considered to be unimolecular reactions that directly involve the atoms of an aromatic ring structure and require no other reagent than heat. These reactions can be categorized in two major types: one that involves a complete and permanent skeletal reorganization (isomerization), and one in which the atoms are scrambled but no net change in the aromatic ring occurs (automerization). The general reaction schemes of the two types are illustrated in Figure.

This class of reactions was uncovered through studies on the automerization of naphthalene as well as the isomerization of unsubstituted azulene, to naphthalene. Research on thermal rearrangements of aromatic hydrocarbons has since been expanded to isomerizations and automerizations of benzene and polycyclic aromatic hydrocarbons.

Mechanisms

Automerizations

The first proposed mechanism for a thermal rearrangement of an aromatic compound was for the automerization of naphthalene. It was suggested that the rearrangement of naphthalene occurred due to reversibility of the isomerization of azulene to naphthalene. This mechanism would therefore involve an azulene intermediate and is depicted below:

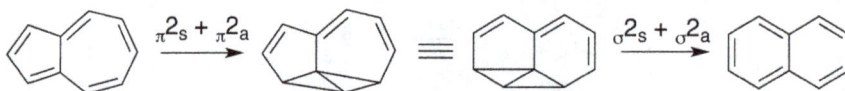

Subsequent work showed that the isomerization of azulene to naphthalene is not readily reversible (the free energy of a naphthalene to azulene isomerization was too high - approximately 90 kcal/mol). A new reaction mechanism was suggested that involved a carbene intermediate and consecutive 1,2-hydrogen and 1,2-carbon shifts across the same C-C bond but in opposite directions. This is currently the preferred mechanism and is as follows:

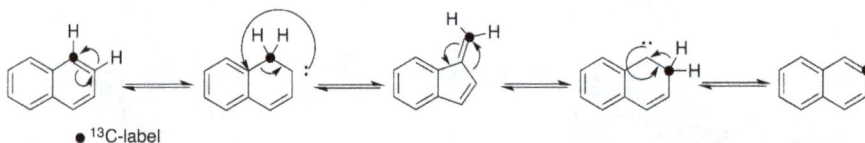

$\bullet$ ^{13}C-label

Isomerizations

The isomerization of unsubstituted azulene to naphthalene was the first reported thermal transformation of an aromatic hydrocarbon, and has consequently been the most widely studied rearrangement. However, the following mechanisms are generalized to all thermal isomerizations of aromatic hydrocarbons. Many mechanisms have been suggested for this isomerization, yet none have been unequivocally determined as the only correct mechanism. Five mechanisms were originally considered: a reversible ring-closure mechanism, which is shown above, a norcaradiene-vinylidene mechanism, a diradical mechanism, a methylene walk mechanism, and a spiran mechanism. It was quickly determined that the reversible ring-closure mechanism was inaccurate, and it was later decided that there must be multiple reaction pathways occurring simultaneously. This was widely accepted, as at such high temperatures, one mechanism would have to be substantially energetically favored over the others to be occurring alone. Energetic studies displayed similar activation energies for all possible mechanisms.

Four mechanisms for thermal isomerizations have been proposed: a dyotropic mechanism, a diradical mechanism, and two benzene ring contraction mechanisms; a 1,2-carbon shift to a carbene preceding a 1,2-hydrogen shift, and a 1-2-hydrogen shift to a carbene followed by a 1,2-carbon shift. The dyotropic mechanism involves concerted 1,2-shifts as displayed below. Electronic studies show this mechanism to be unlikely, but it must still be considered a viable mechanism as it has not yet been disproven.

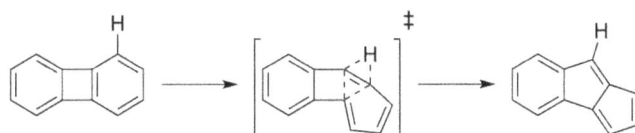

The diradical mechanism has been supported by kinetic studies performed on the reaction, which have revealed that the reaction is not truly unimolecular, as it is most likely initiated by hydrogen addition from another gas-phase species. However, the reaction still obeys first-order kinetics, which is a classical characteristic of radical chain reactions. A mechanistic rational for the thermal rearrangement of azulene to naphthalene is included below. Homolysis of the weakest bond in azulene occurs, followed by a hydrogen shift and ring closure so as to retain the aromaticity of the molecule.

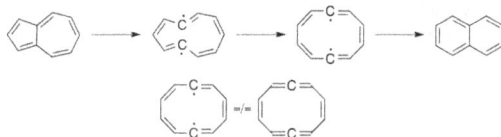

Benzene ring contractions are the last two mechanisms that have been suggested, and they are currently the preferred mechanisms. These reaction mechanisms proceed through the lowest free energy transition states compared to the diradical and dyotropic mechanisms. The difference between the two ring contractions is minute however, so it has not been determined which is favored over the other. Both mechanisms are shown as follows for the ring contraction of biphenylene:

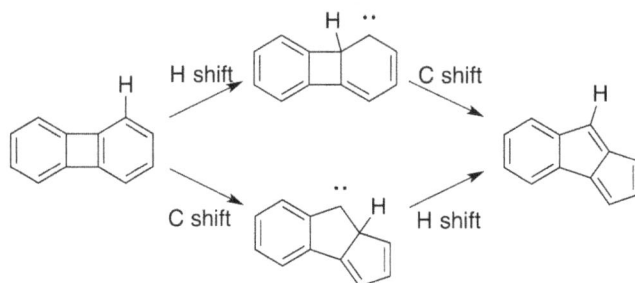

The first involves a 1,2-hydrogen shift to a carbene followed by a 1,2-carbon shift on the same C-C bond but in opposite directions. The second differs from the first only by the order of the 1,2-shifts, with the 1,2-carbon shift preceding the 1,2-hydrogen shift.

The four described mechanisms would all result in the isomerization from azulene to naphthalene. Kinetic data and ^{13}C-labeling have been used to elucidate the correct mechanism, and have led organic chemists to believe that one of the benzene ring contractions is the most likely mechanism through which these isomerizations of aromatic hydrocarbons occur.

History

Indications of thermal rearrangements of aromatic hydrocarbons were first noted in the early 20th century by natural products chemists who were working with sesquiterpenes. At the time, they noticed the automerization of a substituted azulene shown below, but no further structural or mechanistic investigations were made.

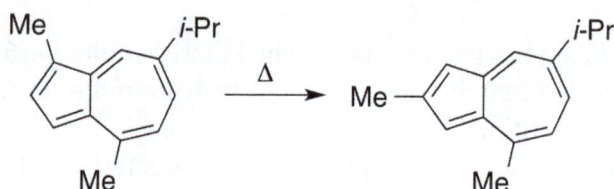

The oldest characterized thermal rearrangement of an aromatic compound was that of the isomerization of azulene to naphthalene by Heilbronner et al. in 1947. Since then, many other isomerizations have been recorded, however the rearrangement of azulene to naphthalene has received the most attention. Likewise, since the characterization of the automerization of naphthalene by Scott in 1977, similar atom scramblings of other aromatic hydrocarbons such as pyrene, azulene, benz[a]anthracene and even benzene have been described. While the existence of these reactions has been confirmed, the isomerization and automerization mechanisms remain unknown.

Reaction Conditions and Flash Vacuum Pyrolysis

Thermal rearrangements of aromatic hydrocarbons are generally carried out through flash vacuum pyrolysis (FVP). In a typical FVP apparatus, a sample is sublimed under high vacuum (0.1-1.0 mmHg), heated in the range of 500-1100 °C by an electric furnace as it passes through a horizontal quartz tube, and collected in a cold trap. Sample is carried through the apparatus by nitrogen carrier gas.

FVP has numerous limitations:

- First, it requires a slow rate of sublimation to minimize bimolecular reactions in the gas phase, limiting the amount of material that can be reacted in a given amount of time.

- Second, the high temperatures used in FVP often lead to reactant or product degradation. Combined, these first two limitations restrict FVP yields to the range of 25-30%.

- Third, the high temperatures used in FVP do not allow for the presence of functional groups, thereby limiting possible products.

- Fourth, as FVP is a gas-phase process, difficulties are frequently encountered when scaling above the milligram level.

- Fifth, the FVP synthesis of strained systems mandates temperatures exceeding 1100 °C, which can lead to the degradation and softening of the expensive quartz apparati.

Possible Applications

Thermal rearrangements of aromatic hydrocarbons have been shown to be important in areas of chemical research and industry including fullerene synthesis, materials applications, and the formation of soot in combustion. Thermal rearrangements of aceanthrylene and acephenanthrylene can yield fluoranthene, an important species in syntheses of corannulene and fullerenes that proceed through additional internal rearrangements.

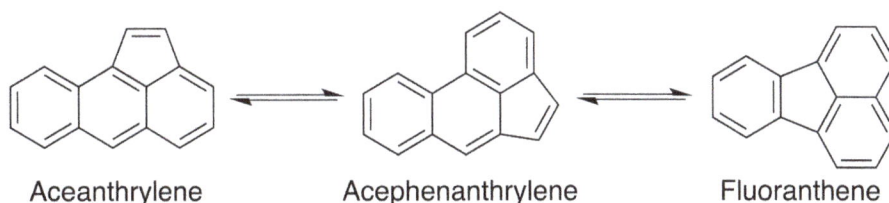

| Aceanthrylene | Acephenanthrylene | Fluoranthene |

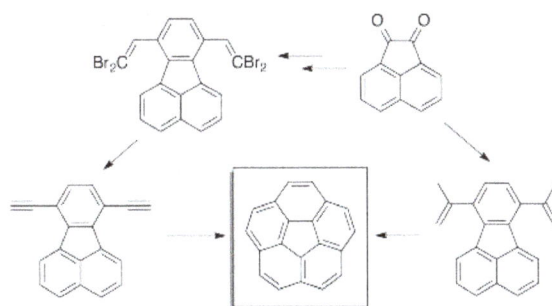

Scott's flash vacuum pyrolysis syntheses of corannulene.

Many of the polycyclic aromatic hydrocarbons known to be tumorigenic or mutagenic are found in atmospheric aerosols, which is connected to the thermal rearrangement of polycyclic aromatic hydrocarbons in fast soot formation during combustion.

Hofmann–Martius Rearrangement

The Hofmann–Martius rearrangement in organic chemistry is a rearrangement reaction converting an N-alkylated aniline to the corresponding ortho and / or para aryl-al-kylated aniline. The reaction requires heat, and the catalyst is an acid like hydrochloric acid.

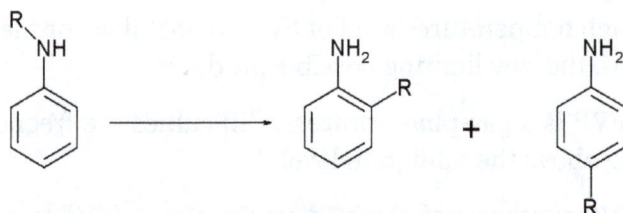

When the catalyst is a metal halide the reaction is also called the Reilly–Hickinbottom rearrangement.

The reaction is also known to work for aryl ethers and two conceptually related reactions are the Fries rearrangement and the Fischer–Hepp rearrangement. Its reaction mechanism centers around dissociation of the reactant with the positively charged organic residue R attacking the aniline ring in a Friedel–Crafts alkylation.

In one study this rearrangement was applied to a 3-N(CH$_3$)(C$_6$H$_5$)-2-oxindole:

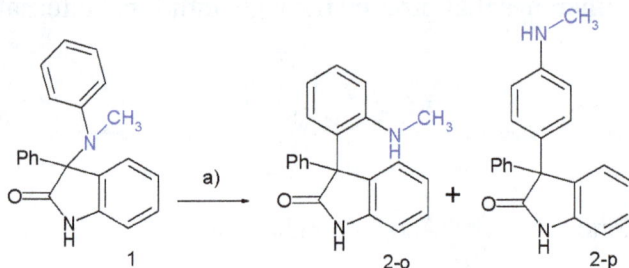

Fischer–Hepp Rearrangement

The Fischer–Hepp rearrangement is a rearrangement reaction in which an aromatic N-nitroso or nitrosamine converts to a carbon nitroso compound:

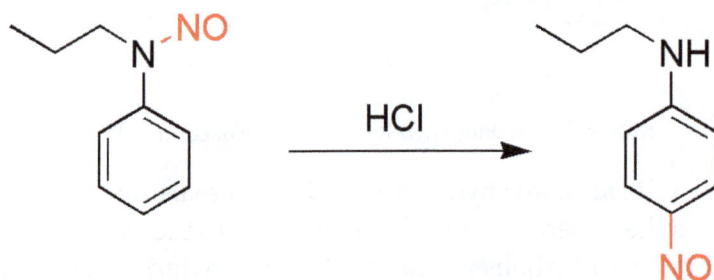

This organic reaction was first described by the German chemist Otto Philipp Fischer (1852–1932) and Eduard Hepp (June 11, 1851 – June 18, 1917) in 1886, and is of importance because para-NO secondary anilines cannot be prepared in a direct reaction.

The rearrangement reaction takes place by reacting the nitrosamine precursor with hydrochloric acid. The chemical yield is generally good under these conditions, but often much poorer if a different acid is used. The exact reaction mechanism is unknown but there is evidence suggesting an intramolecular reaction.

Bamberger Rearrangement

The Bamberger rearrangement is the chemical reaction of N-phenylhydroxylamines with strong aqueous acid, which will rearrange to give 4-aminophenols. It is named for the German chemist Eugen Bamberger (1857–1932).

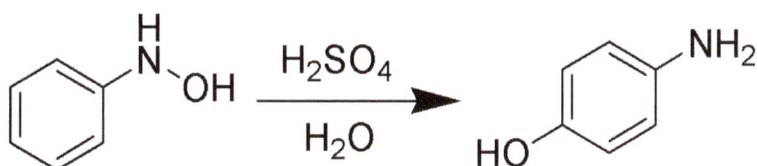

N-Phenylhydroxylamines are typically synthesized from nitrobenzenes by reduction using rhodium or zinc.

Reaction Mechanism

The mechanism of the Bamberger rearrangement proceeds from the monoprotonation of N-phenylhydroxylamine 1. N-protonation 2 is favored, but unproductive. O-protonation 3 can form the nitrenium ion 4, which can react with nucleophiles (H_2O) to form the desired 4-aminophenol 5.

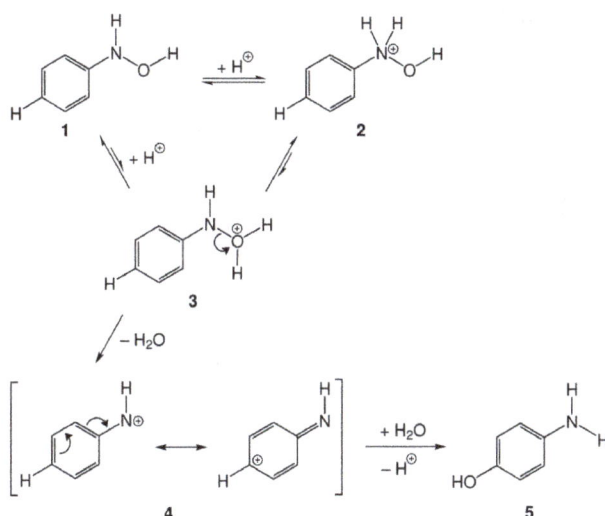

Fries Rearrangement

The Fries rearrangement, named for the German chemist Karl Theophil Fries, is a re-arrangement reaction of a phenolic ester to a hydroxy aryl ketone by catalysis of Lewis acids.

It involves migration of an acyl group of phenol ester to the aryl ring. The reaction is ortho and para selective and one of the two products can be favoured by changing reaction conditions, such as temperature and solvent.

Mechanism

Despite many efforts, a definitive reaction mechanism for the Fries rearrangement has not been determined. Evidence for inter- and intramolecular mechanisms have been obtained by crossover experiments with mixed reactants. Reaction progress is not dependent on solvent or substrate. A widely accepted mechanism involves a carbocation intermediate.

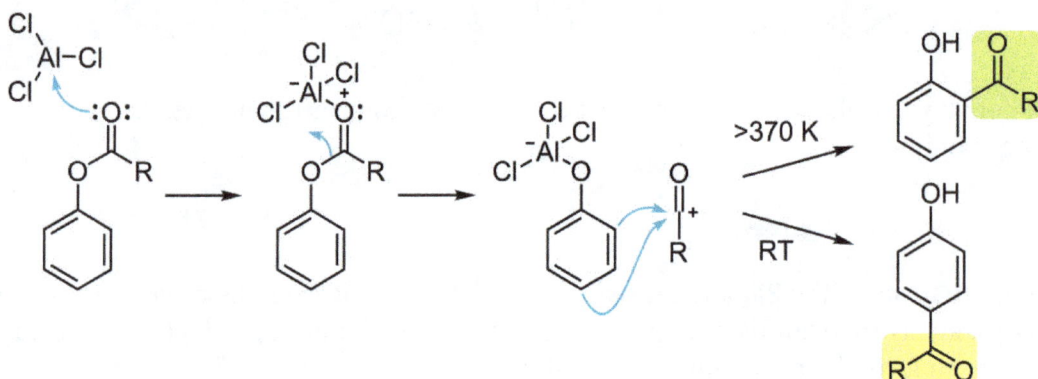

In the first reaction step a Lewis acid for instance aluminium chloride $AlCl_3$ co-ordinates to the carbonyl oxygen atom of the acyl group. This oxygen atom is more electron rich than the phenolic oxygen atom and is the preferred Lewis base. This interaction polarizes the bond between the acyl residue and the phenolic oxygen atom and the aluminium chloride group rearranges to the phenolic oxygen atom. This generates a free acylium carbocation which reacts in a classical electrophilic aromatic substitution with the aromatic ring. The abstracted proton is released as hydrochloric acid where the chlorine is derived from aluminium chloride. The orientation of the substitution reaction is temperature dependent. A low reaction temperature favors para substitution and with high temperatures the ortho product prevails, this can be rationalised as exhibiting classic Thermodynamic versus kinetic reaction control as the ortho product can form a more stable bidentate complex with the Aluminium. Formation of the ortho product is also favoured in non-polar solvents; as the solvent polarity increases, the ratio of the para product also increases.

Scope

Phenols react to form esters instead of hydroxyarylketones when reacted with acyl halides under Friedel-Crafts acylation conditions. Therefore, this reaction is of industrial importance for the synthesis of hydroxyarylketones, which are important intermediates for several pharmaceuticals. As an alternative to aluminium chloride, other Lewis acids such as boron trifluoride and bismuth triflate or strong protic acids such as hydrogen fluoride and methanesulfonic acid can also be used. In order to avoid the use of these corrosive and environmentally unfriendly catalysts altogether research into alternative heterogeneous catalysts is actively pursued.

Limits

In all instances only esters can be used with stable acyl components that can withstand the harsh conditions of the Fries rearrangement. If the aromatic or the acyl component is heavily substituted then the chemical yield will drop due to steric constraints. Deactivating meta-directing groups on the benzene group will also have an adverse effect as can be expected for a Friedel–Crafts acylation.

Photo-Fries Rearrangement

In addition to the ordinary thermal phenyl ester reaction a so-called photochemical Photo-Fries rearrangement exists that involves a radical reaction mechanism. This reaction is also possible with deactivating substituents on the aromatic group. Because the yields are low this procedure is not used in commercial production. However, photo-Fries rearrangement may occur naturally, for example when a plastic bottle made of polycarbonate (PC) is exposed to the sun, particularly to UV light at a wavelength of about 310 nm, if the plastic has been heated to 40 degrees Celsius or above (as might occur in a car with windows closed on a hot summer day). In this case, photolysis of the ester groups would lead to leaching of phthalate from the plastic.

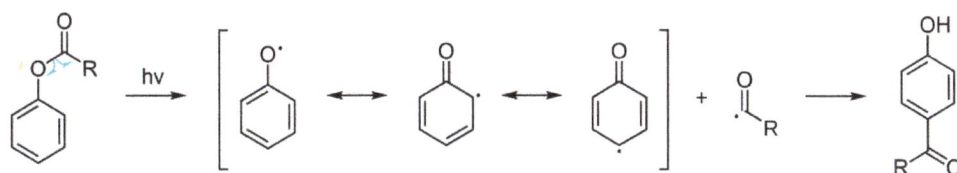

Anionic Fries Rearrangment

In addition to Lewis acid and photo-catalysed Fries rearrangements, there also exists an anionic Fries rearrangement. In this reaction, the aryl ester undergoes ortho-metalation with a strong base, which then rearranges in a nucleophilic attack mechanism.

References

- Schrock, R. R.; Meakin, P. (1974). "Pentamethyl complexes of niobium and tantalum". J. Am. Chem. Soc. 96 (16): 5288–5290. doi:10.1021/ja00823a064

- Sambasivarao Kotha; Kuldeep Singh (2007). "Cross-enyne and ring-closing metathesis cascade: A building-block approach suitable for diversity-oriented synthesis of densely functionalized macroheterocycles with amino acid scaffolds". European Journal of Organic Chemistry. 2007 (35): 5909–5916. doi:10.1002/ejoc.200700744

- March, Jerry (1985), Advanced Organic Chemistry: Reactions, Mechanisms, and Structure (3rd ed.), New York: Wiley, ISBN 0-471-85472-7

- "Dow Corning and Elevance Announce Partnership to Market Naturally Derived Ingredients in Personal Care Applications" (Press release). Elevance Renewable Sciences. 9 September 2008. Retrieved 19 January 2012

- Schrock, R; Rocklage, Scott; Wengrovius, Jeffrey; Rupprecht, Gregory; Fellmann, Jere (1980). "Preparation and characterization of active niobium, tantalum and tungsten metathesis catalysts". Journal of Molecular Catalysis. 8 (1–3): 73–83. doi:10.1016/0304-5102(80)87006-4

- Everett, Wallis; Lane, John (1946). "The Hofmann Reaction". Organic Reactions. 3 (7): 267–306. ISBN 9780471005285. doi:10.1002/0471264180.or003.07

- Whitmore, Frank C. (1932). "The common basis of molecular rearrangements". J. Am. Chem. Soc. 54 (8): 3274–3283. doi:10.1021/ja01347a037

- Mohan, Ram S.; Monk, Keith A. (1999). "The Hofmann Rearrangement Using Household Bleach: Synthesis of 3-Nitroaniline". Journal of Chemical Education. 76 (12): 1717. doi:10.1021/ed076p1717

- Shioiri, Takayuki (1991). "Degradation Reactions". Comprehensive Organic Synthesis. 6: 795–828. ISBN 9780080359298. doi:10.1016/B978-0-08-052349-1.00172-4

- Über Triphenylmethyl-peroxyd. Ein Beitrag zur Chemie der freien Radikale Wieland, H. Chem. Ber. 1911, 44, 2550–2556. doi:10.1002/cber.19110440380

- Baumgarten, Henry; Smith, Howard; Staklis, Andris (1975). "Reactions of amines. XVIII. Oxidative rearrangement of amides with lead tetraacetate". The Journal of Organic Chemistry. 40 (24): 3554–3561. doi:10.1021/jo00912a019

- Plagens, Andreas; Laue, Thomas M. (2005). Named organic reactions (2nd ed.). Chichester: John Wiley & Sons. ISBN 0-470-01041-X

- Astruc D. (2005). "The metathesis reactions: from a historical perspective to recent developments" (abstract). New J. Chem. 29 (1): 42–56. doi:10.1039/b412198h

- Nyfeler, Erich; Renaud, Philippe (24 May 2006). "Intramolecular Schmidt Reaction: Applications in Natural Product Synthesis". CHIMIA International Journal for Chemistry. 60 (5): 276–284.doi:10.2533/000942906777674714

- Shioiri, Takayuki (1991). "Degradation Reactions". Comprehensive Organic Synthesis. 6: 795–828. ISBN 9780080359298. doi:10.1016/B978-0-08-052349-1.00172-4

- Milligan, Gregory L.; Mossman, Craig J.; Aube, Jeffrey (October 1995). "Intramolecular Schmidt Reactions of Alkyl Azides with Ketones: Scope and Stereochemical Studies". Journal of the American Chemical Society. 117 (42): 10449–10459. doi:10.1021/ja00147a006

- Strategic Applications of Named Reactions in Organic Synthesis Laszlo Kurti, Barbara Czako Academic Press (4 March, 2005) ISBN 0-12-429785-4

- R.R. Schrock (1986). "High-oxidation-state molybdenum and tungsten alkylidene complexes". Acc. Chem. Res. 19 (11): 342. doi:10.1021/ar00131a003

- Wrobleski, Aaron; Sahasrabudhe, Kiran; Aubé, Jeffrey (May 2004). "Asymmetric Total Synthesis of Dendrobatid Alkaloids: Preparation of Indolizidine 251F and Its 3-Desmethyl Analogue Using an Intramolecular Schmidt Reaction Strategy". Journal of the American Chemical Society. 126 (17): 5475–5481. PMID 15113219. doi:10.1021/ja0320018

- March, Jerry (1985), Advanced Organic Chemistry: Reactions, Mechanisms, and Structure (3rd ed.), New York: Wiley, ISBN 0-471-85472-7

Permissions

All chapters in this book are published with permission under the Creative Commons Attribution Share Alike License or equivalent. Every chapter published in this book has been scrutinized by our experts. Their significance has been extensively debated. The topics covered herein carry significant information for a comprehensive understanding. They may even be implemented as practical applications or may be referred to as a beginning point for further studies.

We would like to thank the editorial team for lending their expertise to make the book truly unique. They have played a crucial role in the development of this book. Without their invaluable contributions this book wouldn't have been possible. They have made vital efforts to compile up to date information on the varied aspects of this subject to make this book a valuable addition to the collection of many professionals and students.

This book was conceptualized with the vision of imparting up-to-date and integrated information in this field. To ensure the same, a matchless editorial board was set up. Every individual on the board went through rigorous rounds of assessment to prove their worth. After which they invested a large part of their time researching and compiling the most relevant data for our readers.

The editorial board has been involved in producing this book since its inception. They have spent rigorous hours researching and exploring the diverse topics which have resulted in the successful publishing of this book. They have passed on their knowledge of decades through this book. To expedite this challenging task, the publisher supported the team at every step. A small team of assistant editors was also appointed to further simplify the editing procedure and attain best results for the readers.

Apart from the editorial board, the designing team has also invested a significant amount of their time in understanding the subject and creating the most relevant covers. They scrutinized every image to scout for the most suitable representation of the subject and create an appropriate cover for the book.

The publishing team has been an ardent support to the editorial, designing and production team. Their endless efforts to recruit the best for this project, has resulted in the accomplishment of this book. They are a veteran in the field of academics and their pool of knowledge is as vast as their experience in printing. Their expertise and guidance has proved useful at every step. Their uncompromising quality standards have made this book an exceptional effort. Their encouragement from time to time has been an inspiration for everyone.

The publisher and the editorial board hope that this book will prove to be a valuable piece of knowledge for students, practitioners and scholars across the globe.

Index

www.ingramcontent.com/pod-product-compliance
Lightning Source LLC
Chambersburg PA
CBHW061947190326

41458CB00009B/2807